R. McLaughlin

# *Wetland Landscape Characterization*

## JOHN G. LYON

Ann Arbor Press

Copyright © 2001 Sleeping Bear Press
All rights reserved

This book represents information obtained from authentic and highly regarded sources. Reprinted material is quoted with permission, and sources are indicated. A wide variety of references are listed. Every reasonable effort has been made to give reliable data and information, but the author and publisher cannot assume responsibility for the validity of all material or for the consequences of their use.

Neither this book nor any part may be reproduced or transmitted in any form or by any means, electronic or mechanical, including photocopying, microfilming, and recording, or by any information storage or retrieval system, without permission in writing from the publisher.

Ann Arbor Press
310 North Main Street
P.O. Box 20
Chelsea, MI 48118
www.sleepingbearpress.com
Ann Arbor Press is an imprint of Sleeping Bear Press

Printed and bound in the United States.
10 9 8 7 6 5 4 3 2 1



Library of Congress Cataloging-in-Publication Data

Lyon, J. G. (John G.)
   Wetland landscape characterization : techniques and applications for GIS, mapping, remote sensing, and image analysis / by John Grimson Lyon.

     p. cm.
Includes bibliographical references (p.   ) and index.
   ISBN 1-57504-121-9
   1. Wetlands—Remote sensing. 2. Geographic information systems. 3. Ecological mapping. 4. Wetland ecology. I. Title.
   QH87.3 .L96 2000
   333.91'8—dc21

00-009454

# About the Author

Lyon was interested early on in wetlands and other moderately disturbed systems as places to evaluate the condition of native vegetation communities. This interest was honed in his youthful wanderings in the mountains and alpine systems of the Pacific Northwest, California, Nevada, and Alaska. Systematic study of wetlands in undergraduate work at Reed College in his native Oregon, and graduate work at the University of Michigan yielded bachelor's and master's theses and a doctoral dissertation devoted to wetlands and other natural systems. Eighteen years as a faculty member and ultimately a full Professor of Civil Engineering and Natural Resources at Ohio State University were further devoted to scholarly pursuits of these interests. A body of work included remote sensor and GIS evaluations of wetlands, development of field methods for wetlands identification, and other efforts on soil, agriculture, riverine, and Laurentian Great Lakes systems.

# Preface

The issue of identifying wetlands, quantifying their change over time, and characterizing the influences of nature and humans on them is a difficult one. Wetlands are a mix of terrestrial and aquatic systems that create a unique condition. Long of secondary interest in the minds of the public, scientists, and engineers, they have been evaluated on a systematic basis in detail only as of late. Now, a number of people are seeking knowledge of wetlands and techniques to better characterize, monitor, and maintain these unique features in the landscape. To that end, *Wetland Landscape Characterization* was written.

# Contents

**1 INTRODUCTION**   1

Wetland Landscape Characterization   1
The Wetland Issue   3

**2 EVALUATIONS FOR WETLAND LANDSCAPE CHARACTERIZATION AND RISK ASSESSMENT**   13

**3 FIELD AND REGIONAL WETLAND MAPPING METHODS**   19

Wetlands Inventory and Identification   20
Analysis of Regional Wetlands   21
Regional or Large Area Assessments   21
Local Area or Small Area Assessments   22
Hydric Soils   24
Presence of Wetland Plants   25
Wetland Soil Types   26
Wetland Hydrology   27
Field Methods   29
Surveying   35

## 4  IDENTIFICATION AND MONITORING — 39

- Information Needs — 39
- Monitoring — 40
- Wetland Ecosystem and Receptor Monitoring — 40
- Detection and Identification of Features — 40
- Management — 42
- How to Characterize Change in Features — 43
- Features of Interest — 45
- Information for Planning — 47
- Indicators of Water — 48
- General Condition of River and Stream Wetlands — 49
- Other Risks and Impacts — 49
- Identification of Habitat for Conservation — 51
- Human-Related Features and Wetland Mitigation — 52
- Hazardous Waste — 53

## 5  IMAGERY AND INTERPRETATIONS — 55

- Remote Sensing — 56
- Photointerpretation — 57
- Interpretations — 60
- Historical Aerial Photographs — 63
- Characterization of Vegetation — 67
- Soil Characterizations — 67
- Water Resource Characterization — 68
- Wetland Classification — 69
- Creation of Aerial Photo Land Cover Products — 70
- Detection of Change Methods — 70
- Detection of Change Using Satellite Data — 74
- Preprocessing of Satellite Data — 78
- The NALC Experience and Change Detection Procedures — 79
- Steps in Change Detection and Data Processing — 80
- Change Detection through Vegetation Index Differencing — 81

Change Detection through Principal Components Analysis — 82
Postcategorization Change Detection — 83
Training Set Development — 83
Categorization — 84
Land Cover and Classification System — 85
Assessment of Accuracy — 85
Applications to Wildlife Habitat Quality Evaluations — 87

## 6 GIS APPLICATIONS — 89

Background on Geographic Information Systems (GIS) — 89
The Utility of GIS — 90
Data Sources and Their Application — 92
Wetland GIS Applications — 95
Aerial Photo Analyses of Historical Wetlands — 96
Use of Remote Sensor Data and GIS — 97
Applications to Marine Coastal Wetlands — 97
GIS Analyses of Large Lakes and Lacustrine Wetlands — 100
GIS for Water Quality Assessments — 101
Lake Water Resource Applications — 103
Related Water Resource Applications — 104

## 7 EVALUATIONS OF ACCURACY — 105

Assessments of Data Quality — 105
Quality Assurance/Quality Control (QA/QC) Issues and Approaches — 105
Assessments of Data Accuracy — 106
Data Analysis — 107

**BIBLIOGRAPHY** — 109

**INDEX** — 131

# 1 Introduction

## Wetland Landscape Characterization

The last ten years have witnessed the development and refinement of a paradigm to address large areas or landscapes, and their combination of biota, and physical and chemical processes (Figure 1.1). The paradigm of the landscape sciences encompasses the disciplines and subdisciplines necessary to address the characteristics and ecology of the landscape scale. Landscape science incorporates the concepts of landscape ecology and landscape characterization.

Landscape ecology is the mix of biological, chemical, and physical processes that characterize the earth and water at the landscape scale (Figure 1.2). Landscape characterization is the combination of methods and approaches that allow the characterization of earth and water resources and the

*Figure 1.1.* The combination of water and vegetation and standing water over soils make a wetland, such as this canal area of the Great Dismal Swamp in Virginia.

*Figure 1.2.* Movement of water such as this tidal flow causes areas to exhibit partial wetland characteristics during the day and perhaps less clear characteristics at other times at Cook Inlet, Alaska.

processes that drive the systems. Together, landscape ecology and landscape characterization provide for the theoretical and practical characterization, evaluation, and prediction of landscape resources and the dynamics of their processes.

The landscape sciences represent the culmination of years of study and theoretical development. The result is a set of concepts that can be used to formulate the problems, execute their study in nature, help in the parameterization and simulation of real processes in a mathematical manner, and the prediction of future trends and the risk that is posed by natural and anthropogenic stresses. An approach which is currently used in the landscape sciences is that of environmental risk assessment.

The development of risk assessment concepts was fostered by a parallel approach of human risk assessment. Environmental risk assessment approaches have a number of component parts that guide the study of landscape science. Environmental risk assessment itself is broken into parts that are related to natural causes of risk or ecological risk, and human-influenced or anthropogenic risks.

Risk assessment involves the characterization of the risk, and how it may influence or expose the resources or receptors to the risk. The exposure of a resources or receptor to a natural or human-induced stress creates the problem. To evaluate the potential risk or conduct risk assessment involves the determination of endpoints or assessment criteria. The endpoints come in two forms, that of assessment and measurable endpoints. Often the endpoints of interest are difficult to evaluate, so that surrogate variables or measurable endpoints are used.

***Figure 1.3.*** *Ebb and flow of water on the surface of soils often is enough to keep soils saturated beneath the surface in the absence of standing water in some areas of the rocky intertidal wetlands of the Oregon coast.*

The risk can be determined by the characterization of exposures and ecological effects to receptors by stressors. Once determined, it becomes important to evaluate the concern through risk characterization and to address the correction of the problems through risk management.

Important considerations in using risk assessment in landscape evaluations are the spatial and temporal characteristics (Figure 1.3). Ecosystems vary over time and space and the risk assessment paradigm incorporates these variables, as does the use of remote sensor data and GIS technologies.

Wetland landscape characterization (Figure 1.4) and wetland landscape ecology encompass the risk assessment and characterization concepts necessary to measure, model, and predict the current, historical, and future status of wetlands and related ecosystems. Here, the concepts necessary to characterize (Figure 1.5) and evaluate the wetlands and related habitats are addressed, along with the methods to conduct these efforts.

## The Wetland Issue

Wetlands and the issue of their management and preservation now engage the attention of the public. Children learn early on that the particular mixture of water and land results in a

*Figure 1.4.* Obligate wetland plants such as cattail often help identify the presence of wetlands.

*Figure 1.5.* Many wetland areas appear problematic as compared to most people's ideal such as in Figure 1.4. Areas with few soils and few plants but with anaerobic soils fit the definition of a wetland, such as these areas on the shore of Vancouver Island, British Columbia.

Introduction

unique ecosystem that they can explore (Figure 1.6). Adults see the value of wetlands as home to the plants and animals they love. All seek to recreate near or in wetlands as they are often found close to home, or near their vacation spots.

There is also a correlated level of interest in maintaining the wetlands that are currently present, and adding to the current quantities of wetlands. This is in addition to the charge of regulatory and management agencies. The public need and governmental oversight makes for a good future for maintaining wetlands.

The most characteristic wetland areas can be recognized by many people. A real problem is that many wetlands are temporary, or look very similar to terrestrial or aquatic ecosystems and may be mistaken as such (Figure 1.7). Many people are comfortable with their knowledge of wetlands, and often believe they know how a given type of wetland should appear (Figure 1.8).

*Figure 1.6. Wetlands and ponds supply recreation and learning opportunities that many people take advantage of, including summer program students in Ohio.*

The major concern is that decisions on management of land areas are made sometimes with a poor understanding of all the different wetland characteristics and functions (Figure 1.9). Many landowners and others often identify wetland areas in a different manner than do regulators or experts at delineation. These regulators and experts are experienced at identifying the less-than-apparent wetland types. Wherever land areas have the potential to be waterlogged or flooded, you are looking at a potential wetland area (Figure 1.10). Depending on local and regional interests, activities in and around these potential wetlands may create harm and may come under intensive scrutiny.

Unlike localized environmental concerns, such as buried steel tanks, hazardous waste, and endangered species, wetlands can be found in a variety of climates and landscapes throughout the U.S. The abundance of wetlands depends on the local hydrology, geomor-

*Figure 1.7. Many areas experience floods or aperiodic periods of standing water and saturated soils, and may have wetland functions for short periods of time such as rain-flooded areas of the Santa Cruz River near Tucson, Arizona.*

*Figure 1.8. The pond with lily pads fits many peoples' definition of a wetland as well as governmental statutes and regulations.*

Introduction

*Figure 1.9.* Less typical wetlands are hard for the lay person to identify, including these salt cedar covered shorelines of Lake Mead in Arizona.

*Figure 1.10.* Spring runoff of snowmelt waters creates many wetland areas that may not be present later in the summer, such as these near Valdez, Alaska.

*Figure 1.11. Wetlands persist in cold areas due to warm tidal water such as these in Prince William Sound near Valdez, Alaska.*

phology, and other natural conditions. Hence, local conditions dictate how frequently they are encountered in many activities.

The widespread distribution of wetlands holds the promise of the resource being found on lands slated for change in land cover, and/or new management practices. Where such ecosystem risks may occur, personnel are working and naturally they will encounter wetlands in the course of practicing their profession. Where development and wetlands co-exist, there are often potential or realized risks, which must be addressed due to federal, state, and local regulations and oversight. Hence, interested parties must pay attention to the potential for risks to wetlands, and it is critical to know the location and variety of wetland ecosystems found in the landscape.

There are also many opportunities to work with wetland ecosystems, to avoid any disturbance, and make the improvement in the numbers of wetlands part of the overall plan of land and water management.

Wetlands are land areas that are periodically flooded or covered with water (Figure 1.11). It is the presence of water at or near the soil surface for more than a few weeks during the growing season that may help to create many wetland conditions. The water slows diffusion of oxygen into the soil and to plant roots. Lack of oxygen or anaerobic conditions cause major changes in the soil chemistry. Only certain "wetland plants" have adapted to life in these harsh conditions. Their adaptations allow them to use available soil nutrients, and they exhibit a variety of physical and physiological adaptations to grow in the absence of available oxygen.

The combination of anaerobic and waterlogged soils, the presence of "wetland" plants, low-lying topography, and other conditions help to create a different land cover type called wetlands. These characteristics and conditions are also used to define and identify wetlands.

Different types of wetlands have been created by hydrological and topographical conditions. This has a lot to do with the variety of water bodies or sources of water associated with the wetland. For example, wetlands adjacent to rivers take on the characteristics of the riparian and riverine conditions. Wetlands on the shore of lakes have many hydrological characteristics that are driven by the lake system. Wetland areas on marine coasts have coastal characteristics and are also influenced by the varying salinity concentrations from open ocean, coastal ocean and neighboring estuarine waters. Hence, the hydrology of a given area is important to the characteristics and conditions, the functions of wetlands, and ultimately to their identification.

For example, the variability of temporary or ephemeral wetlands makes them particularly hard to distinguish. These areas may only display wetland function for a little over two weeks during the growing season. Yet they are vital to the ecology of most ecosystems, such as the desert or semidesert ecosystems in the western United States, Canada, and Mexico.

One central priority of the wetland issue is the definition of a wetland. In particular, we are interested in the definition of a jurisdictional wetland (JW) (Lyon, 1993). It is defined in the U.S. Army Corps of Engineers 1987 wetland manual as, "Those areas that are inundated or saturated by surface or ground water at a frequency and duration sufficient to support, and that under normal circumstances do support, a prevalence of vegetation typically adapted for life in saturated soil conditions. Wetlands generally include swamps, marshes, bogs, and similar areas."

A jurisdictional wetland is defined in the field. It must exhibit a dominance of wetland plants, soils subject to waterlogging or hydric soils, and indicators of wetland hydrology. These three individual criteria must be addressed, and areas exhibiting all three parameters are deemed jurisdictional wetlands.

Due to the scale of many inventories of wetland ecosystems, general wetlands (GW) or potential jurisdictional (PJW) wetlands (Lyon, 1993; Lyon and McCarthy, 1995) are addressed here in detail. These wetlands have one or more of the federal criteria wetland characteristics and may also be jurisdictional wetlands (JW), but include adjacent non-wetlands or converted wetlands. Definitions and methods for their identification and assessment of risk are addressed.

Personnel work in and about wetland areas on a frequent basis. In land development, wetlands are encountered by personnel in evaluations of property. They are encountered in the surveying and photogrammetry work necessary to produce a large-scale topographic map of the property, or in a number of other activities. The presence or absence of wetlands may be important in any work on property.

The current level of interest and the spatial variability of the resource argue for an organized approach to their characterization and evaluation. The variability and widespread presence of wetlands, and their patch-like distribution, make it necessary to employ appropriate technologies that allow evaluations. Scientists, engineers, and the public are challenged to employ knowledge and experience to address the characteristics of these resources, and analyze their conditions.

The distributed nature of wetlands requires that their locations be identified for thoughtful management, preservation, and evaluation of risks. There are several ways to locate and inventory general wetlands, and assess the risks posed by stressors. It is necessary to accomplish these goals related to wetland inventory, and to do so in a manner that matches the needs of the community, and the financial and human resources that are available. There are also several different levels of detail that can be collected about wetland areas, in order to evaluate risks associated with human activities. These activities are all part of wetland landscape characterization and risk assessment.

Wetland landscape characterizations are required for a number of reasons. Water quality and quantity sources drive the ecosystem, and are a unifying concern. The wetlands themselves are an amalgam of processes. The mix yields a variety of functions for all lifeforms, and a variety of interactions with human activities. To bring together the processes, functions, and uses necessitates approaches and methods that are inherently spatial, and that incorporate concepts of environmental risk and environmental health.

To better understand and manage wetland landscapes creates a desire to examine current capabilities for measurements, inventory, and modeling. Gone are the days when simple facts and emotion ran the dialogue. In their place are new tools to augment traditional examples so as to supply quantitative data.

We now can measure wetland and related ecosystems with great numbers of samples, and rely on computers to store this wealth of empirical data. Further processing with simple or complex algorithms based on theory and practice allows for a quantitative and accurate view. Empirical and simulation results tested by assessments of accuracy now pace the dialogue, and power decisions with facts and predictions based on quantitative, empirical evidence.

To measure wetland and related landscapes of this breadth in character, the difficulty of collecting data becomes obvious (Figure 1.12). The heart of analysis is the quantity and variety of data. Most research efforts in a laboratory or conducted on experimental plots seek to control all variables but one, and study different levels of the selected variable of interest or the large, forcing function or stressors. In macroscale or landscape-scale analyses, the effort to control a given variable or a few variables is hard and potentially expensive. Often landscape-scale studies utilize survey sampling techniques (Cochran, 1977; Congalton and Green, 1998) and large-size samples to address variables. This can be a successful approach where the typical empirical approach may not be feasible.

To obtain the best quality and variety of data on the landscape scale, survey sampling and other appropriate measurement and data organization technologies are often used. Collection of data from a distance or remote sensing has demonstrated value for a number of

Introduction

*Figure 1.12.* More typical wetlands can be found wherever the combination of water, soils, and vegetation prevail, such as this log pond in Western Oregon.

applications. The variables of interest may be directly measured over large areas using uniform data collection methodologies. Indirect variable or surrogate variables may also be measured to take advantage of the capabilities of remote sensing. Examples of direct measurements would be the inventory of general wetland (GW) types. Indirect measurements would be the location of different general wetland types, their presence and absence, their interspersion, their juxtaposition and perhaps fragmentation (Lyon, 1983; Robinson et al., 1992) as compared to the known habitat requirements of biota too small to resolve with a remote sensor.

Copious amounts of data are usually appreciated in landscape-scale evaluations. Yet these data must be organized with human interpretation or computer analyses. The advent of spatial databases or Geographic Information Systems (GIS) allows the storage of data in quantity. The spatially based storage of data as coordinates or vectors, or as grid-cells in a raster matrix, makes implicit the spatial location of the data. GIS allows retrievable storage and spatial fidelity, and can accommodate as much data as are necessary. The current efforts and future efforts are devoted to modeling of processes using GIS and remote sensor imagery systems.

To address methods to evaluate the presence and characteristics of wetlands is the goal of this book. A variety of techniques, methods, and approaches have been brought together to address this need for information. Most of these have been evaluated by the author in the field or laboratory and found to be useful.

# Evaluations for Wetland Landscape Characterization and Risk Assessment

The identification and inventory of general wetland types can be potentially valuable (Figure 2.1). Of greater value is the inventory of wetlands over relatively long time periods, and use of mapping and GIS technologies for evaluations and monitoring, and for assessments of risk.

Early work on multiple year evaluations of wetlands focused on collecting multiple date data sets. This allowed examination of trends in numbers of general wetlands, and allowed collection of wetland data from different hydrologic conditions or hydroperiods. Due to the variability of weather and of storage of water in shallow depressions and in ponds, the extent of wetlands may change over time. In marine systems, the

*Figure 2.1. Submergent wetlands are found beneath the surface of the water, and often have a cover of plants such as these in the foreground of this picture from a southern Michigan lake.*

*Figure 2.2. Emergent wetlands are composed of plant species which mostly pierce the water and emerge into the air, such as these emergent wetlands in southern Alaska.*

variability of tides and influence of storm events requires the use of tidal gauge data and/or multiple measurements of the location of wetlands as related to the daily and seasonal tidal cycles and storms.

Once multiple dates of wetland quantities are collected, a number of scenarios can be evaluated to fix causative factors and processes. The characterization of probable causative factors or stressors for wetland changes can begin from the assembled long period record. General wetlands (GW) may change in both extent and the variety of wetland types. An increase in the storage of water on a lake may push the location of the water's edge inland, and conversely the decrease in storage may move the water edge lakeward (Jaworski et al., 1979; Lyon, 1981, 1993). These changes may become stable over a year-to-year period and cause emergent wetlands, for example, to change to shrub-scrub wetlands with decreasing storage, or to change to unconsolidated bottom during times of high water storage (Roller, 1977; Cowardin et al., 1979; Carter, 1982; Williams and Lyon, 1991).

Early analyses of the change in general wetlands focused on year-to-year increases or decreases based on lake storage (Jaworski et al., 1979). Lake systems are largely influenced by storage as a stressor on the long-term, and by wind or air pressure induced changes over the short-term (Lyon, 1981; Williams and Lyon, 1997), called seiches. This has allowed the evaluation of water level fluctuations over time in the absence of tides characteristic of marine environments (Lyon and Drobney, 1984; Bukata et al., 1987; Williams and Lyon, 1991).

These data collection methods have helped facilitate the analysis of trends in quantities and varieties of wetlands (Figure 2.2). Large area evaluations of wetlands have been made with imagery acquired from aircraft and spacecraft.

These airborne or space sensors offer several advantages (Figure 2.3). The repetitious satellite coverage which began in 1972 allows for multiple date assessments of general wetlands (Lunetta et al., 1993). The resolution of satellite sensors ranges from approximately 80 meters to 10 meters over time. Now, 1 to 5 meter data are regularly available and more commercial sensors supplying finer resolution spatially and spectrally are coming on-line. The availability of 1 meter visible and 3 to 5 meter multi-spectral data will be of great assistance. If the resolution is suitable to the applications of interest, then these data are potentially useful.

The common spectral bandwidths and resolution include the visible and near infrared portions of the spectrum. A number of sensors also incorporate the middle infrared and thermal infrared portions which also have a variety of uses in studies of wetland, lake, or water resources.

*Figure 2.3.* The aerial view allows one to locate the wetland ecosystems of interest and to assess the surrounding area for natural and human features such as this shot of the central U.S.

An important application of satellite remote sensor data has been the development of regional or continental data sets (Lunetta et al., 1993; Shaw et al., 1993) which address wetlands. Of particular note are continental efforts aimed at wetlands and other land cover types (Dobson and Bright, 1993), and use of remote sensor data for evaluation of wetland and other wildlife receptors or habitats for presence and absence of resources (Browder et al.,1989; Scott et al., 1993).

Early work focused on demonstrating the capabilities of satellite remote sensing and use of nonvisible light measurements of wetlands (Roller, 1977; Lyon, 1978; Lyon, 1979;

## Wetland Landscape Characterization

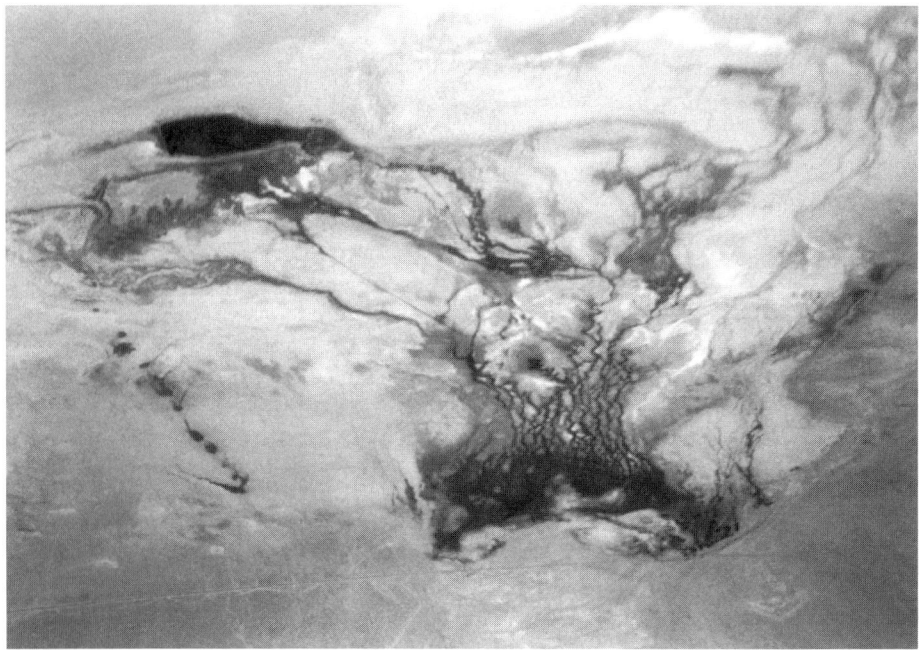

*Figure 2.4.* From above, water appears dark-toned as it reflects little light, and bare or partially vegetated uplands appear light-toned in this picture of Wyoming.

Lyon, 1980; Carter, 1982; Lee and Lunetta, 1995). Initially, users were challenged by the complexity of freshwater wetlands as they were often composed of many plants species and were influenced by different hydrologic and hydroperiod factors (Figure 2.4). Marine coastal wetlands were somewhat easier to address because some wetland types were dominated by one or two species of plants. These examples included Spartina, Salicornia, and Mangrove wetlands.

Recent work has demonstrated the maturation of technologies for general wetland (GW) identification and inventory. Particularly interesting and valuable efforts have been conducted in the southern United States and other areas (Welch et al., 1988; Jensen et al., 1992; Remillard and Welch, 1992; Welch et al., 1992; Jensen et al., 1993; Remillard and Welch, 1993; Ramsey and Jensen, 1995; Ramsey and Jensen, 1996). These efforts have made great use of remote sensing and advanced the use of GIS for storage and analysis of results.

Later work demonstrated the utility of imagery for evaluations of freshwater wetlands. Multiple dates of aerial photographs provide good data on general wetland quantities and varieties, and on the hydrologic conditions that prevail (Jaworski et al., 1979; Lyon, 1981; Carter, 1982; Lyon and Drobney, 1984; USEPA, 1991; Lampman, 1993). Early work demonstrated that many dates of photographs and the resulting photointerpretation efforts yielded quantitative data of historical and current conditions (Jaworski et al., 1979; Lyon, 1981; Lyon and Drobney, 1984; Williams and Lyon, 1991; Lyon and Greene, 1992). These applications have helped to develop theories and models of the underlying causes of change in wetland types (Jaworski et al., 1979; Lyon et al., 1986; Bukata et al., 1987; McKee et al., 1988; Jensen et al., 1992; Jensen et al., 1993).

A number of wetland inventories have been conducted over the years to supply information for local and state management of wetland resources. Early efforts were conducted in New Jersey, Wisconsin, Ohio, and various local areas to meet management goals and objectives. The current condition is that many entities have developed inventory data for their own purposes. These efforts have made use of aerial photographs, field evaluations, satellite remote sensor data, and GIS technologies to complete the application and later manage the information. The groups have been diverse, including cities, townships, conservation and environmental organizations, and corporate entities including utilities and the forest industry.

Several groups have demonstrated the value of satellite and aerial remote sensing in wetland inventory. Ducks Unlimited has completed a number of satellite remote sensor evaluations of wetlands in the Dakotas and in Manitoba and Saskatchewan. The Nature Conservancy's natural heritage program has evaluated a variety of resources for preservation on a state-by-state basis. These programs have made good use of remote sensor data in identifying and inventorying resources that are of great interest to the public.

The Ohio Wetlands Inventory has made use of a combination of remote sensor and other data (Yi et al., 1994). Sources included Landsat Thematic Mapper (TM) data sets, existing GIS databases of soils and other sources of information (Channell, 1989) called the Ohio Capability Analysis Program (OCAP), U.S. Geological Survey Digital Line Graph (DLG) map theme data, and field visits (Schaal, 1995). To identify general wetland areas, the near and middle infrared bands of the TM sensor were used to sense wet soils (Thenkabail et al., 1992; Lee and Lunetta, 1993; Thenkabail et al., 1994). This has been found to be particularly useful in soil moisture analyses and analysis of plant moisture conditions (Van Deventer et al., 1997; Lunett and Balogh, 1999).

The National Wetland Inventory (NWI) conducted by the U.S. Fish and Wildlife Service has developed maps of general wetlands for most of the United States. This effort is discussed later, but it represents a large governmental effort to provide one-time mapping of general wetlands.

Other important efforts have been devoted to elucidating models or paradigms to evaluate processes and their interactions as they influence ecology at the landscape scale. The ecological risk assessment paradigm has been developed over the years to provide a framework to measure and model the landscape. The advantages of the ecological risk assessment and landscape characterization paradigms are multifold, as the latter feeds the former. Hence, the links between ecological risk assessment and landscape characterization are close and are becoming even closer.

# 3
# Field and Regional Wetland Mapping Methods

In the absence of regional and local knowledge of wetland ecosystems and receptors, it has become desirable for interested parties to assess these resources. A very good way to educate parties as to the presence of wetlands is to quantify the variety of wetlands and produce mapping products. These sorts of activities facilitate the making of decisions related to the thoughtful management of the landscape, and avoidance of or disturbance of wetlands.

There are a number of compelling reasons to identify and map wetlands on a countywide, local, or other basis. The value of wetland ecosystems and receptors is well known and it is desirable to preserve these systems for their utility and function. A local, county or regional assessment can provide information and allow for informed decisions on land use and zoning, and help characterize the current number and variety of wetlands.

Oftentimes these wetland areas are adjacent to the site of community development and must necessarily be evaluated. Due to a variety of interests and legislative mandates it is prudent to identify these areas and quantify their size.

Depending on the goals of a given community project, a set of methods may be suggested to characterize wetlands, and their location on the landscape. Whether the goal is to inventory general wetlands (GW) or to delineate jurisdictional wetlands (JW), the activity must be planned and approaches must be selected.

The scale of the mapping product or inventory often plays a prominent role in selection of methods. If the goal is to identify, delineate, and map a wetland of approximately one-half acre in size, the methods are going to be different than if the goal is to inventory general wetland types in a given region of more than ten thousand acres. A general or potential jurisdictional (PJW) wetland is defined as having one or more of the three indicators of wetlands (USACE, 1987; Lyon, 1993). A general wetland can be detected by visits to the field, from aerial photographs or from satellite remote sensing data or other techniques (Lyon, 1993; Lyon and McCarthy, 1995; Ward and Elliot, 1995). A general wetland is distinct from a jurisdictional wetland.

urisdictional wetland is defined by procedures largely conducted in the field using prevailing federal and/or local procedures such as the U.S. Army Corps of Engineers manual (USACE, 1987), and is potentially subject to regulatory oversight.

A general wetland may or may not be a jurisdictional wetland, but a general wetland can also be different based on the criteria above. The definition and identification of a general wetland allows for assessment and inventory of wetlands over large areas using remote sensing and other technologies, and a certain amount of field evaluation.

There is a need for these definitions, as groups must be able to inventory general wetlands using a variety of techniques, including fieldwork. Jurisdictional wetlands necessarily need to have a high proportion of fieldwork to determine their presence, and to meet the requirements of regulatory standards. Here, general wetland types are addressed due to issues of scale.

Many groups have been able to map the general wetland plant community types from aerial photos. From the regional perspective, such a determination and an identification of types using an available wetland categorization system would be feasible and valuable using aerial photos. Hence, aerial photos or other remote sensor data are often used in these efforts.

## Wetlands Inventory and Identification

The detection of general wetlands can be made in the field, or by use of a combination of fieldwork and aerial photographs or other remote sensor technologies (Carter, 1982; USACE, 1987; FICWD, 1989; Carter, 1990; Lyon, 1993).

Initial efforts to develop an inventory involves the identification of wetlands from a given field evaluation and/or sets of aerial photographs. Due to the fundamental nature of this large area or regional scale of inventory, the emphasis is on detection of general wetlands and their identification as to type using a categorization or classification system such as the USGS Anderson (1976) system or the U.S. Fish and Wildlife Service National Wetlands Inventory (NWI, Cowardin et al., 1979) system (Carter, 1982; Butera, 1983).

Maps are often the starting point in any inventory or analysis, and often are used in the presentation of results in reports or for presentations at meetings. U.S. Geological Survey (USGS) and other original source maps may be used to obtain a variety of information at medium scales. They can be used to collect: point location information; topographic contour information; cultural or planimetric details such as roads, waterways, and dwellings; or public land ownership boundaries. The maps can also provide the starting point to form a variety of land cover themes or layers in a GIS.

One should also make use of Soil Survey documents (USDA, 1962, 1975, 1991). Some of these data are available as digital files. These digital files of soils may be employed for analysis if they are appropriate to the scale of the inventory effort (SCS, 1992; Lytle, 1993).

Engineers often have large-scale maps available for site planning and design. The scale is commonly one inch equal to one hundred feet ($1'' = 100'$) or one inch equal to two hundred feet (1″ = 200′). The contour interval is often one foot of elevation for engineering design, and a two-foot interval for planning purposes. These products are especially useful in hydrological analyses, and they are created through surveying or a combination of surveying and photogrammetric technologies. These products may also include details as to where wetlands may be found. However, the wetland or marsh symbol has been applied to the map by the stereocompilation operator, where he or she deemed there to be a wetland. These areas should not be used as evidence of wetlands because the operator usually has had no training in wetlands photointepretation or delineation. Also, the field checks have not necessarily been made as to the accuracy of the location and size of the wetland details supplied on the map.

## Analysis of Regional Wetlands

To identify resources in a county or multiple county area it is best to use a combination of maps and aerial photos. The photos will be both historical and of recent origin. It is desirable to use a variety of available photos from archives, because one can accumulate historical data on wetlands: during different times; under different hydrological conditions; during leaves-off and leaves-on conditions; and to take advantage of different film types that may reveal more information than black and white photos alone.

If maps showing general wetlands are available, they can be employed as additional data for the inventory. These may be available from different state or local sources, and one should try to identify any products which may be available.

The U.S. Fish and Wildlife National Wetland Inventory (NWI) has been an early and ongoing effort to develop inventory information on national, general wetlands (Wilen, 1990; Lyon, 1993). This work has involved the use of aerial photographs, soil survey information, field visits, and GIS technologies to develop a database throughout the United States. It has employed an agreed-upon and uniform classification system (Cowardin et al., 1979), and has made use of national personnel, contract personnel, and state personnel thus far. A disclaimer is printed on each map product to provide the user with guidance as to the appropriate level of detail and accuracy that is supplied.

The NWI data are available in most locations in the United States, or will be soon. They generally represent a one-time analysis of general wetlands. For GIS purposes, the NWI supplies certain general wetland information which can be of value in various applications including inventory of general wetlands.

## Regional or Large Area Assessments

The intent of a large area assessment is to identify and map general wetland areas in county-sized or larger areas, and create wetland maps potentially useful for planning activities.

The steps in the project may include the following:

(a) If possible, collect new photos of the site at relatively large scale in the fall and/or spring. These aerial photos provide for two different moisture conditions, and record current land use or cover. Black-and-white film can be used during leaves-off conditions. Fly color infrared (CIR) film during leaves-on conditions to document the growing season foliage. CIR products and growing season data are very valuable, but the foliage may obscure wetland hydrology and hydric soil conditions. Hence, images from several seasons are very desirable (Lyon, 1993; Lyon and McCarthy, 1995; Lunetta and Balogh, 1999).

(b) Secure existing photographs from a variety of sources including local and national archives. Use these photos to help identify wetlands under the variable hydrological conditions experienced over the years.

(c) Collect information from aerial photographs, the county Soil Survey, topographic maps, and any other pertinent data sources and identify general wetland boundaries on an overlay of the available maps. Initially, areas will be mapped on 1:24,000 scale quadrangles using the fall aerial photos and other data.

(d) These quadrangle overlay products and additional photos from the spring and growing season will be used to further identify general wetland areas on the maps.

(e) Integrate other data sources such as watershed and drainage maps to help qualify wetland resource areas that are close to or on important water bodies or stream/river waterways that are potentially of interest.

(f) Fieldwork is valuable to determine the accuracy of mapped general wetland areas. The expert can make corrections. Establish and conduct an accuracy assessment experiment to determine the quality of the final products (Congalton and Green, 1998).

(g) Combine products and digitize them into permanent files to develop final maps.

(h) Provide these products and a report summarizing the methods of production. Describe the characteristics of the products and their accuracy.

(i) Provide an opportunity for evaluation of the products by clients or officials.

(j) Incorporate fieldwork and criticisms, and produce final products identifying general wetland areas.

## Local Area or Small Area Assessments

Currently, no large-scale map products exist or will exist in the future that supply information on general wetlands or potential jurisdictional wetlands, and/or jurisdictional wetlands in most local areas. Completion of these maps would greatly facilitate planning, thoughtful development, and ecosystem preservation activities within given localities. Hence, it is nec-

## Field and Regional Wetland Mapping Methods

*Figure 3.1.* There are a variety of field indicators of wetlands and the three criteria used to identify wetlands. Here the presence of wetland plants, flooded or saturated soils, and wetland hydrology are evident at Killdeer Plains, Ohio.

essary to undertake specific mapping and fieldwork efforts to characterize the presence and variety of wetlands on the large scale or local scale.

This approach described here will focus on efforts at the local level and the different wetlands found there. The result of the effort will be an enhanced posture to deal with this difficult-to-quantify resource, and an attempt to dispel the great confusion accompanying this issue when detailed wetland maps are unavailable.

The finding of a regulatory, jurisdictional wetland is based on criteria set forth in the interagency, "Wetlands Delineation Manual" (USACE, 1987) and supporting documents, regulations, statutes, and guidance letters from USACE.

An area is considered a regulatory, jurisdictional wetland only if all three wetland indicator criterion requirements are met. The criteria include a determination as to: (a) whether the soils are considered hydric or waterlogged, (b) whether the soils show indicators of wetland hydrologic conditions associated with flooding or ponding of water and/or saturation of soils, and (c) whether 50% or more of the plants found growing on the site are truly wetland plants.

Areas which fail to satisfy one or more of these conditions or criterion are not considered jurisdictional wetlands.

The phenomena that cause a wetland criterion or parameter to be found are varied, as are the indicators used to make a regulatory wetland assessment. It is desirable to address the three criteria in detail (Figure 3.1).

## Hydric Soils

Soils subject to flooding, saturation, or ponding of water for more than one to two weeks per year will often demonstrate hydric soil characteristics. Water saturation or ponding of water over soils causes the loss of oxygen due to chemical and biological oxygen demands. In a period of a few days or less, these demands will use the available oxygen in soils including the root zone area of plants.

The dissolution of oxygen in water and migration into the soil and root zone is very slow and proceeds at an unacceptable rate compared to the needs of upland plants to respire. The absence of oxygen in the root zone area will commonly cause the mortality of upland plants during the growing season. This is because the plant roots need to respire or use oxygen to metabolize sugar for their life-supporting, energy requirements.

*Figure 3.2. Wetland hydrology can be inferred by the presence of silt or clay deposits or flotsam, shown here from water transport processes on Lake Michigan.*

An important criterion to identify jurisdictional wetlands is the presence of an abundance of plants adapted to grow in wetland or waterlogged soils. These plants are adapted to grow in specific environments and compete with other plants for nutrients and light. Areas which experience periodic flooding or ponding of water are commonly populated by wetland plants as opposed to upland plants (Figure 3.2).

One competitive advantage is the capability to grow and reproduce in very low oxygen or anaerobic soils. While this is a very stressful environment, many plants have the capability to exploit this advantage and grow in exclusion of other plants not used to low concentrations of air. Simply put in ecological terms, wetland plants exploit this stressful environment to grow and reproduce, thereby excluding upland plants and/or other less competitive plants and thereby providing themselves a "home" or niche.

Wetland plants have a variety of adaptations and structures or chemical processes which allow them to withstand short or long periods of water-saturated soils or ponding of water. These adaptations to supply oxygen to the root zone include structures such as aerenchyma, or alternative chemical processes such as the formation of malic acid.

## Presence of Wetland Plants

The identity of plants that can withstand oxygen deficiencies for short or long periods are known, and they are referred to as wetland plants. The characteristic distribution of plant species in wet areas is rated in one of five categories, and by the different regions of the country. Their identities and abundance are used to characterize whether an area is a wetland for the plant criteria.

A list of plant species and their affinity for wetland and upland conditions has been published and is referenced here as Reed et al. (1988). The system provides that certain plants are found most commonly in upland ecosystems (UPL). Many types of plants are present in wetland areas less than 33% of the time they are located, and are called facultative upland plants (FACU). Other plants occur in wetland areas with a frequency of between 34–66% of the times they are located on a given site. They also occur in upland or in nonwetland areas, and are labeled facultative (FAC) wetland species. Plants which have a special affinity or capability to live in wetland areas occur there with a frequency between 67–99% of the time they are located in a given site, and are termed facultative wetland plants (FACW). Wetland-loving species that occur almost all the time in wetlands with a frequency of greater than 99% of the time on a given site are called obligatory (OBL) wetland plants.

The list of plants sometimes includes a suffix of "+" or "-" to indicate the plant's affinity to or frequency of being found in a wetland category.

In the field, personnel need to identify the plant species that are the most common or dominant in a given wetland and surroundings. These common plants need to be evaluated as to their frequency or abundance by percentage. The evaluation of frequency can be made for each dominant plant species in each layer of vegetation including the tree, vine, shrub, and ground layers.

The dominant plants need to be identified at the species level so that they can be evaluated as to their wetland indicator status and relative abundance. The composition of the community of wetland plants is made by enumerating the FAC, FACW, and OBL plants found in a given location. One would identify the plant species that are the most common or dominant, and estimate their abundance by percentage. Such a determination is made for each layer of vegetation including the tree, shrub, and ground layers.

The area is found to be a wetland for this individual criterion when the total abundance of FAC, FACW, and OBL plants exceeds 50% of the total plants found on the site, as integrated across all the layers or strata.

For identification purposes one can collect plants in the field and freeze or refrigerate them until there is time to identify them in the office. Once an individual is thoroughly familiar with the dominant types, they may be described in the field without collecting plants at each individual sample point. Knowledge of plants and a variety of plant identification books should be used for keying plants to individual species types. Consider drying select plants in a plant press of layered cardboard secured by straps. These plants may be useful later as a record of what was identified and what was found on a given site.

## Wetland Soil Types

*Figure 3.3. A good method to use in evaluating soils for wetland indicators is to dig a hole approximately 18 inches deep, and view subsequent seepage and/or standing water.*

Soils often are subjected to flooding, ponding, and/or saturation by water. Soils with these conditions that continue over time are characterized as hydric. It is possible to identify hydric soils by examining the surface and subsurface characteristics of the soils. These characteristics include the presence of dark soil colors associated with the absence of oxygen, rust mottling associated with varying conditions of oxygen abundance, dark soil color or chroma resulting from chemical reactions that occur in the absence of oxygen, and other characteristics (Figure 3.3).

It is also useful to identify the soil as to name or type and characteristics as mapped and described in the USDA Soil Surveys. The U.S. Department of Agriculture (USDA) has developed a list of soils which often display hydric soil conditions. Soils found in an area and found on the USDA list of hydric types are potentially wetland soil types and will often exhibit hydric conditions.

County Soil Surveys are available from the county offices of the Natural Resource Conservation Service (NRCS), formerly the Soil Conservation Service (SCS), of the USDA. The

Field and Regional Wetland Mapping Methods

*Figure 3.4.* After evaluating a grid node for wetland characteristics it is good to mark or flag the spot so as to be able to recover the location to check the work at a later time.

nearest office and phone number can be found in the government section of the phone book. The USDA Hydric Soils List can be obtained from the USDA-NRCS, Washington, D.C. (National Technical Committee for Hydric Soils Criteria for Hydric Soils, USDA, SCS, 1991) or the U.S. Government Printing Office.

Evaluation of a Soil Survey for soils on the USDA Hydric Soils List is a necessary first step in any effort. If said soils are present, it will often be incumbent upon the land owner to determine if a jurisdictional wetland is present by analysis of the site for the criteria (Figure 3.4).

## Wetland Hydrology

Land areas that are subject to flooding, ponding, or saturation of water generally display surface characteristics indicative of the actions of water. Surface indicators of water include flood-borne, water-deposited debris (Figures 3.2 and 3.5). Layers of clay or silt particles on the ground or on plants are also indicative of standing or flowing waters. These indicators supply detail as to past water-related events. They may be used in the field and are descriptive in character.

Surface indicators may be observed in the field as one walks the site. Subsurface characteristics can be identified by evaluating soil conditions. This may be done by shoveling

*Figure 3.5.* Several indicators of wetland conditions are shown here. They include multiple trunk trees, flotsam, and standing water.

*Figure 3.6.* Expanded tree boles are an indicator of saturated soils.

out a hole over 18 inches deep or by using a soil probe to remove soil cores for visual examination. Soil color characteristics including hue, value, and chroma can be identified using a Munsell color chart for soils and for hydric soils. Further information can be taken from the USACE 1987 Manual or USDA, 1975, Soil Taxonomy book.

Next observe whether a water table or seepage is encountered within 18 inches of the soil surface. This is a strong indicator of wetland hydrological conditions, an important measure of soil hydrology. These measurements are best performed during the growing season (USACE, 1987).

Another method of determining general wetland (GW) potential jurisdictional wetland (PJW) hydrological conditions is to evaluate the proximity of a given property to rivers, streams, gullies, and nearby wetlands of significance. This can be performed without a field visit and may be completed by using available maps or aerial photos. Field visits are also useful to further identify the characteristics of the waters.

The presence of a navigable waterway may also be determined at the same time. Maps of the navigable waterways are available from the Regulatory Branch of the U.S. Army Corps of Engineers District or Area Office. All these characteristics can be important in the assessment of jurisdictional or general wetlands (Figure 3.6).

## Field Methods

As outlined above, the wetland expert must identify possible jurisdictional or regulatory wetland areas using Soil Survey maps and evaluation of surface and subsurface indicators of hydric soils, indicators of waterlogged soils and/or flooding associated with wetland hydrology, and from evaluations of dominant plant species and the relative percentage of those plants that are commonly found in wetlands.

To identify general or other wetlands in a given area it is best to use a combination of existing data, reference materials, and field evaluation procedures. The following activities may be beneficial in evaluations of regulatory wetlands for the "routine" level of evaluation as described in USACE (1987).

The steps in the preliminary or "routine" analysis would include:

(a) Collecting available documents that characterize the site including the county Soil Survey, work done by others, and any engineering-style large-scale topographic maps. Integrate this information mentally, and/or incorporate the mapping details into a base map. Include the boundaries of soils on the site that are found on the USDA Hydric Soils List, boundaries of any water course, and boundaries of general wetlands from NWI maps.

(b) Using the available information and any maps or base maps, visit the site and walk the entire area. Evaluate all parts of the site, and pay particular attention to any area identified as a possible jurisdictional or general wetland due to soil types, hydrology, vegetation, and topography that is low-lying. Evaluate all water

*Figure 3.7. Flooding can greatly expand the extent of saturated soils, but waters often recede during the growing season as seen along the Muskingum River in Ohio.*

bodies including rivers, ponds, lakes, streams, creeks, gullies, wet spots, and obvious wetlands (Figure 3.7). Look for any anomalous conditions of soil, bedrock, or hydrology that might cause a wetland to form. It is important to examine hydric soil areas as recorded in the Soil Survey.

(c) Integrate the field-work, aerial photographs, the county Soil Survey, and any other pertinent data and identify possible jurisdictional wetland boundaries on an overlay of the topographic map. The desirable minimum mapping unit would be one-tenth to one-hundredth of an acre. This forms the basis for a preliminary or "Routine" (in 1987 Manual terminology) evaluation of wetland location and quantities. Using a digitizer make a preliminary estimate of the total potential or general wetlands.

(d) Combine information from other data sources, such as watershed and drainage maps, to help characterize wetland resource areas as potentially of regulatory interest.

Following this procedure and steps in the 1987 Manual provides a certain minimum level of information to allow identification of areas which are potential jurisdictional wetlands. In the language of the 1987 Manual, this is a preliminary or Routine level report suitable for identifying the scope of the problem.

***Figure 3.8.*** *Soil color can be evaluated by use of the Munsell color charts.*

If wetlands are detected, and the size of the wetland area exceeds one acre, and the wetlands may be disturbed, a more detailed analysis may be required for a variety of reasons. The next effort is called an "advanced" or comprehensive level evaluation. This product is valuable for making an actual determination of wetland quantity. Such a document can be used to respond to wetland-related questions posed by the public or governmental entities.

To complete an advanced or comprehensive level evaluation it is necessary to:

(a) Develop a grid system of points to sample in the field. A 100 × 100 foot grid mesh has been found to be suitable (other methods are suggested in the 1987 Manual). At each grid-node it is desirable to sample and record data on soils, hydrology, and plants. Results can be listed on the appropriate forms supplied in the 1987 Manual or by vendors who have developed good forms. Resulting information can be used in identifying regulatory wetlands based on the three criteria detailed in the 1987 Manual.

(b) Dig a hole or take a soil probe and evaluate the top 18 inches of soil. This activity includes: watching for water or seepage; examining Munsell color charts for color, value, and chroma; checking for iron oxide mottling; depth of organic matter; manganese reduction products; "gley" or very wet soils; sulfur or methane smell; and other indicators. Be sure to sample each soil layer encountered and record your findings. For general purposes, soils that exhibit standing water, dark Munsell chromas (/2, /1), lots of mottling, a layer or high levels of organic matter, or other evidence of the absence of oxygen and presence of anaerobic conditions, will help to satisfy the hydric criterion (Figure 3.8).

(c) Describe the soil type of each sample point as it compares to the USDA-NRCS soil types presented as an overlay on the topographic map. Record whether sample points have similar or dissimilar soils to those indicated by Soil Survey information. Also look for evidence of hydric soils such as high water table or standing water during the growing season. Such high water table or soil saturated conditions are usually defined as a period of time exceeding five percent of the growing season. The growing season is defined as the period between soil temperatures exceeding 32 degrees Fahrenheit, and this detail can be found in the county Soil Survey.

To establish that the area meets this criterion, one may wish to visit the site many times during the growing season to record the conditions. It may also be desirable to establish a monitoring well of pipe with holes to facilitate measurements.

(d) Examine the soil and wetland landscape for hydrological indicators. At the grid point list any evidence of flood markings and debris flotsam, shallow or adventious root systems, decomposed plants, adjacent river or gulley or stream courses, and any other indicators of wetland hydrology mentioned in the 1987 Manual.

(e) Examine the plants found in the area. The ground, shrub, vine, and tree layers of the existing vegetation must be evaluated by plant species. In the experience of the author, it has been found useful to record the species in the ground layer of vegetation within a circle of radius of approximately 10 feet, the shrub and vine layer within a circle approximately 20 feet in radius, and the tree layer within an approximately 30 foot radius circle of the grid point. It is also necessary to estimate relative abundance of each dominant plant based on the relative prevalence of the given plant species, as compared to the total quantity of vegetation at each sample point.

(f) The plant species data along with their dominance estimates in percentage are used to identify the common plants at the sampling site. The wetland delineation procedure is presented in the 1987 Manual, and it requires an evaluation as to whether 50% of the dominant plants have a high probability of occurring in wetlands. Reed et al. (1988) records the plants species and their "agreed upon" categories of probability of occurrence in wetlands by geographic region.

(g) A given point is a regulatory wetland when it has: (1) hydric soils; (2) indicators of wetland hydrology; and (3) a finding that 50% or more of the dominant plants have a high probability of occurring in a wetland (FACU, FAC, OBL). All three of these parameters must be satisfied. Lack of one or more of the three parameters means that the area is not considered a wetland for regulatory or jurisdictional purposes (Figure 3.9).

(h) It is desirable to plot the locations of the sampling points on the engineering-style large-scale topographic map(s), and to supply the boundaries of any jurisdictional or regulatory wetlands.

It may also be desirable to present the boundaries of any hydric soil encountered on a separate map (Figure 3.10). These boundaries will make a nice contrast with any USDA hydric soil boundaries shown on the Soil Survey.

(i) Employ available sources of information to further understand the characteristics of the area (Figure 3.11). Make use of additional visits to the area to identify any and all resources of interest. Evaluate the delineation effort and any map products and/or flagged boundaries. Also note regional conditions that may influence the significance of a given wetland. Determine whether there are any large wetlands located on neighboring property. Evaluate aerial photos of both historical and of recent origin. It is desirable to use a variety of available photos from archives, because one can accumulate historical data on wetlands during different times, hydrological conditions, and during leaves-off and leaves-on conditions. Take advantage of different film types that may reveal more information than black-and-white photos alone (Figure 3.12).

*Figure 3.9. Oftentimes the indicators of wetlands are found in a local area but not at the same location and in the correct combination. Pictured here is an older Holiday Inn property that has a pool of water and a FACW tree, pin oak, and perhaps periodically saturated soils.*

(j) Set the location of the wetland areas with boundaries that conform to the USACE 1987 Manual. Calculate the size of each wetland in acres.

(k) Allow a period for evaluation of the products by the client. Incorporate

*Figure 3.10.* Oftentimes, certain places in farm fields are too wet to farm and are left out of the management plan. Here is a small wet spot with willow trees in Licking County, Ohio.

comments and criticisms if appropriate, and produce final products identifying potential regulatory wetland areas. These products and a report summarizing the methods used in their production may be submitted as a delineation to the Army Corps of Engineers or other interested party.

This information provides some insight and appreciation of the complexities involved in wetland delineations. This approach can be useful in a number of applications, and provides a good, general model.

On some occasions it is necessary to further define the characteristics and boundaries of jurisdictional wetlands. This may involve the use of advanced level techniques (USACE, 1987). Where wetland boundaries are convoluted, where there are many small wetlands about the site, where there are especially high valued wetlands, there may be a need to used advanced level methods. Many of these methods are discussed here and in other chapters.

It is important to note that many questions about wetlands are mostly legal questions. It is often necessary to seek counsel on these issues. It is also good to note that a scientific and engineering-based report on wetland quantities is a necessary first step in a wetland analysis, and counsel will eventually require this information for guidance purposes.

## Surveying

The widespread distribution of wetlands holds the promise of the resource being found on lands slated for development. Where development occurs, the professional is working and naturally will encounter wetlands in the course of practicing the profession. Where development and wetlands coexist, there are often problems related to federal, state, and local regulatory oversight, and hence the professional must pay attention to the wetland issue.

Professionals work in and about wetland areas on a frequent basis. In land development, wetlands are encountered by surveyors in evaluations of deeds, or in surveying and photogrammetry work necessary to produce a large-scale topographic map of the property. The presence or absence of wetlands may be important in laying out the subdivision, or in other construction related activities.

*Figure 3.11.* Low-altitude aerial photographs provide a good view of sources of water, and general wetland and riparian areas.

At an early stage in site design, the wetland permit requirements will have to be addressed. The wetland issue is addressed as part of the initial environmental report for the property.

Wetlands that are identified as to type often need to be recorded as to location. There are informal methods to accomplish such a task. Formal methods such as a wetland delineation are often employed to provide a permanent record of wetland extent at a certain time. When working in the field at the local scale a number of traditional and new methods can be employed. A very good and accurate method is to determine the location of a wetland as defined by the accepted, current identification methods being employed. The boundaries can be marked with flags or tape, and the boundary can be surveyed by a licensed or registered surveyor (Lyon, 1993; Lyon, 1995).

## Wetland Landscape Characterization

*Figure 3.12. Black-and-white photographs can be acquired with chartered aircraft, and oblique views such as this can show the juxtaposition of natural and human exposures to riverine and wetland systems.*

The scenario is that the wetland delineation expert flags the boundaries. The surveying party locates the boundaries within the parcel, makes field measurements, and recovers monuments as necessary. The products are a map of the delineated wetland boundaries, and a determination of area within the boundary.

Have the wetland delineation expert flag the boundaries of the wetlands that have been identified. This is necessary as the surveyor is merely recording the boundary, and not making decisions about the location of the boundary.

It may be desirable to have the wetland delineator present at the beginning of the survey. This will help assure that all parties know the location of delineated boundaries, and what areas are inside the boundary.

Wetland boundaries are usually more convoluted that human-made boundaries. Hence, surveyors should be prepared to measure a very convoluted boundary or polygon. There will be lots of short distance measurements (Figure 3.13). There will be lots of angles to be turned due to the frequently changing angular location of the boundary.

Why should a survey present such a convoluted polygonal area? This is because permitting decisions are based on total area of wetlands to be filled or discharged into. Hence,

# Field and Regional Wetland Mapping Methods

*Figure 3.13.* Wetlands can be found in many places. Hilltops and hillsides can have wetlands where water ponds and soils stay saturated, such as this reclaimed mine area in Muskingum County, Ohio.

known, precise, and accurate wetland boundary and wetland area estimates are vital to resolving the wetland issue for a given property. As a result, the surveyor should expect variability, short lines, and many measurements. This is due to the variability of the resource, and due to the variability in the boundary delineated by the wetland expert.

The surveyor can potentially provide great assistance. A service can be performed by mapping the delineated boundaries. This can be done while avoiding permitting problems that are neither the concern nor the purvey of surveyors.

Surveying is important in aquatic environments, as these measurements form the basis of many hydrological calculations. The surveyed positions of basins, subbasins, channels, control structures and the like all form important inputs to calculations. These positions may also provide the basis for GIS databases, and for GIS model calculations.

Surveying is also important in the production of photogrammetric products (USACE, 1993; Falkner, 1994). To make photogrammetric calculations in absolute units and to reference measurements absolutely, it is necessary to collect ground-surveyed positions. These positions or ground control points (GCP) allow the stereo model to be tied into absolute ground positions.

Surveying is also used to mark the positions and follow the progress of field construction. These measurements help to assure that the engineering design specifications are met.

The advent of more capable instruments and computers has greatly influenced the contributions of surveying in general, and in particular to landscape characterization. The use of total station instruments has allowed distances to be measured with laser Electronic Distance Measurement (EDM) devices and angles to be turned. Data can be stored directly to an "on-board" computer system to facilitate record keeping, and downloaded later to expedite calculations of positions.

These tools and their application have allowed surveyors to be more efficient and to work with fewer crew members. It has also made the data readily available for computer analyses and production of final products using computer output devices.

Global Positioning Systems (GPS) have now made position information available to most users at minimal cost. This technology has brought surveying to the general public, and has really contributed to the layperson. Outdoor pursuits, general navigation, safety, and science and engineering have all benefited.

GPS is a system of satellites transmitting a microwave signal to earth that can be received by GPS instruments on the surface (Van Sickle, 1996). The signal forms a timing code such that the receiver can measure time of flight at the speed of light, and convert the timing information to a distance. Receiving a number of signals from different satellites allows one to position the receiver with multiple time or distance measurements.

To better define the position, many civilian applications make use of more than one receiver. One receiver is stationary over a known position, and this allows the user to compare the timing signal received from one stationary instrument to another that is moved about to determine positions and elevations. The comparison and use of correction algorithms removes sources of error that can result from the atmosphere or other conditions. This Differential GPS (DGPS) approach can be implemented in a number of manners based on position accuracy and precision requirements.

Wetlands can be located in the field by marking their boundaries as described above. The boundary of large wetland areas can be located using DGPS, and the latitude and longitude or Universal Transverse Mercator (UTM) position of points can be recorded (Norton and Slonecker, 1990; Slonecker et al., 1992). Back in the office, the points can be plotted and connected to complete the polygon describing the boundary of a wetland. This method works well with differential GPS systems and produces accuracy and precision on the order of +/- meters to centimeters in horizontal position. For applications which can utilize that level of accuracy and precision it becomes a very useful approach. If a licensed surveyor and survey is not required by the project objectives, trained staff can develop the product using DGPS. This technique is still useful for obtaining very high accuracies and precision even in the absence of signal disruptions formerly present in civilian systems.

At the regional scale, it may be possible to identify general wetlands with a combination of fieldwork and interpretation of remote sensor data. A number of applications are described below that have utilized this approach. It is an efficient way to generate appropriate scale data for general wetlands.

# 4 *Identification and Monitoring*

## Information Needs

There are a variety of information needs in wetland ecosystem management and landscape characterization. This is due in part to the variety of problems to be addressed. The problems include the traditional examples, and related questions that develop from analysis of the problems.

The traditional problems and questions come from the operational information needs of conservation and environmental groups, and government. These specific data needs may include characteristics such as: wetland feature size and type; wetland exposure and risk characteristics; conditions of a site and facilities or resources; records of management activities; maps; GIS data, and other information.

Now, the domain of these problems includes new requirements beyond the traditional, and the questions they engender. New problems and questions stem from regulatory requirements. These requirements come from a variety of federal, state, and local agencies. These new types of requirements spawn different questions and needs for data.

There is also a new paradigm to address the concerns related to wetland ecosystems. Landscape ecology and landscape characterization incorporate a holistic view of large area systems and their interaction with the variables that influence them. We examine the wetlands and their ecosystem, and how they interact or function as receptors for the exposures generated by stressors. We can integrate the cause and effect of a number of natural or ecosystem risks, and/or the combination of natural and human- induced stressors that constitute environmental risk and risk assessment.

The domain of data needs is now a variable domain. In addition to traditional data needs, new information needs include: data on human factors; data for identification and management of wetland and related ecosystems; cultural features; databases to support management operations and maintenance; and data to support public query and response systems.

# Monitoring

To address the variety of information needs related to planning and management of wetland, aquatic, and terrestrial ecosystems often requires analyses of data over time or monitoring. A monitoring approach should incorporate capabilities to collect and compare data over time. Often these methodologies are termed "change detection" methods and measurements (Lyon et al., 1998).

There are a number of ways to monitor wetland, aquatic, and terrestrial ecosystems over time. Each has a value, and each may be used for certain requirements. Methods that involve imaging measurements have been found useful in these applications. These imaging and other methods include: field inspection and monitoring; aerial photography; aerial videography; and aircraft and satellite imaging (Figures 4.1, 4.2).

A change detection or monitoring methodology can be greatly facilitated by the use of digital imaging data from sensors. Imaging data have the capabilities to supply spatial and spectral data over large areas, and to do so at a potentially lower relative cost than fieldwork and mapping alone.

# Wetland Ecosystem and Receptor Monitoring

The issue of wetland resources monitoring is important for tracking the current, historical, and future quantities and varieties of wetlands. These efforts may range from straightforward assessments of quantity, and change over time, or they may be complex evaluations of quantity and variety of wetlands as influenced by stressors.

# Detection and Identification of Features

A variety of problems can be identified and monitored using remote change detection and monitoring. The fundamentals include the use of two or more images for comparison purposes. Change can be judged by interpretation of the two or more images.

The differences in light reflected from one kind of land cover or land use as compared to another may be used as an indicator that change has occurred (Lyon et al., 1998). The difference in grey tone on black-and-white images, or as color in color or color infrared images, can be interpreted as change over time. This characteristic of change in tone or color can be evaluated from two or more images taken from different times.

The difference in tone or color can be demonstrated. When a forest exists, the tone is dark on black-and-white film and the color is dark green on color film. When houses are constructed in among or nearby the forested wetland or marsh, the tone during construction is light on black-and-white film and the color is brown-white on color film. On color infrared (CIR) films bare soil areas appear blue-green in color and light toned or bright. These

*Figure 4.1.* Moisture conditions change greatly throughout ecosystems. Coastal wetlands can be influenced by long-term differences due to net precipitation in the basin such as these coastal lacustrine wetlands on Lake Michigan.

*Figure 4.2.* Moisture conditions can be influenced by short-term events such as wind or pressure-induced flooding or seiches typical of large lake coastal wetlands on Lake Michigan. This is the same view as in Figure 14.1, but water levels have changed due to long or short-term variations.

*Figure 4.3. Riverine and riparian areas are greatly influenced by runoff which can vary. A riparian forest or marsh may be a wetland area during higher annual runoff conditions due to change in prevailing weather.*

differences are very obvious under most conditions. The differences can be interpreted over time as construction of houses or buildings and as an exposure to wetland receptors.

Other characteristics and conditions of property can be of interest. Indicators of these conditions can be used to identify and locate areas for fieldwork, and contribute to management of wetlands.

## Management

A lot of questions related to monitoring involve encroachment in and around wetland properties. There are a number of imaging and GIS methods to identify the location and the probability that an activity or exposure has occurred.

An activity can be identified by the change in grey tone or color on individual images. Comparison with additional images allows an assessment of conditions, and of whether or not there has been activity or a disruption (Figure 4.3).

Much of this remote capability for identification employs interpretation of image products over time. The use of multiple images can facilitate a thorough, general evaluation of conditions from one time frame to another.

Once a disturbance or exposure has been identified it can be checked in the field. This approach allows for efficient use of personnel, by guiding them to a probable activity. It also represents a good application of capabilities—that of technologies supplying location and identification data, and field personnel checking for activities or exposure, which is a function they do best.

Remote sensor data and systems like GIS will also help in prioritization of field crew work. Information can aid in prioritization of dispatch, and speed their evaluations. The imaging remote sensor data and GIS will supply locations to check. They can also supply data in the form of map sheets, images of features, data on wetland characteristics, and other appropriate data if available.

Methods such as detection of change in wetlands and other land cover areas hold promise. They can be particularly cost-effective as compared to the relative costs of traditional methods of fieldwork and aerial photography. Benefits include the use of satellite and aerial remote sensor data for: detection of change; production of orthophoto-like maps; data inputs to GIS systems; monitoring of resource integrity; evaluation of areas following floods or earthquakes; identification and storage of data concerning cultural and natural resource features; and production of data sheets from GIS for planning and management.

A great deal of research has gone into the development of change detection procedures (Lunetta et al., 1993; Lunetta and Elvidge, 1998; Lyon et al., 1998; and others). The optimal and applicable procedures utilize digital imaging data from two or more dates. The data may come from the same aerial or satellite sensor, or may be from different sources.

The data sets are evaluated for change over time. Initially, one data set is preprocessed to the characteristics of the second data set. The two data sets are also placed in a common mapping framework or mapping projection. Completion of these steps facilitates the comparisons of images.

## How to Characterize Change in Features

It is the change in land cover that acts as an indicator of activities. Methodologies for detection of change have been examined over a period of time. The remote sensor-based technologies offer a great deal of capability for wetland and wetland-related applications.

Indicators of construction are an important identifier of activity, and help to locate areas of new housing or buildings. The new construction areas can be identified by the removal of vegetation and presence of bare earth. The bare earth is commonly made up of mineral soil, which has a high reflectance of light. Bare, mineral earth appears very light-toned on black-and-white film or bright white-brown on color imagery.

On occasion, soils high in organic matter are disturbed. These areas appear to be very dark-toned, and darker than plants in the visible part of the spectrum.

The monitoring of areas will demonstrate construction activities by the change from a vegetated or other land cover to the light-toned bare earth areas of construction. This difference is very simple to detect from satellite or aerial data.

This monitoring and indicator method can identify new construction for a number of years after the fact (Figure 4.4). The change in land cover is distinct for many years because it takes time for native vegetation to repopulate the disturbed soils, or for landscaped plants to mature. Development of structures results in bare earth areas, and subsequent growth of grass, shrubs, and trees is much different from the original land cover and its pattern and tone or color on imagery.

A remote sensor monitoring program can provide this change information. It is general information, but it supplies indicators of large or local activities that have occurred. It still remains for field and office personnel to determine cause, and prioritize the importance of the data and any remedial efforts.

To meet the traditional and new requirements for information in support of industry, conservation and environmental groups, and in support of government requires technologies that supply appropriate data. The new technologies must be flexible to address both traditional and new problems and questions. Digital imaging methods have the necessary capabilities to supply appropriate data.

*Figure 4.4. A number of natural or human-induced conditions can alter wetlands. Growths of kudzu cover native vegetation in the southern U.S., such as this location in the Great Dismal Swamp of Virginia and North Carolina.*

There are additional values associated with imaging data. These data inputs can be used to a greater degree if they are stored and retrieved from computer systems. Imaging data are either gathered as photographs and optically digitized, or are measured directly using a digital instrument on a remote sensing satellite or aircraft.

The appropriate data for monitoring must also be available at a low relative cost. In a monitoring approach, there is the implicit need to acquire image data from two or more periods of time. This often involves at least seasonal and sometimes monthly data acquisitions and the associated costs.

## Identification and Monitoring

Remote sensor technologies provide appropriate methods for monitoring the characteristics of wetlands and adjacent land cover types. Implementation of a monitoring approach requires a combination of input data sets and hardware and software for data processing. The effort is straightforward, and can be employed by a certain company or obtained through contracts with vendors.

The optimal sources of data must also require only limited processing before they can be used for generation of products. The finished product is the input to the monitoring effort, and if the image product requires a great deal of processing it will represent a large cost item. This is particularly true of the multiple images required for monitoring.

Most problems and questions can be addressed using a combination of remote sensor and Geographic Information System (GIS) technologies over time. The resulting data products can be used to determine the presence or absence of existing features. The change in features from one time period to another is a strong indication that an activity or exposure has occurred.

The capabilities of GIS allow for analyses of many data sets. As users employ one or more dates of wetland quantity data and experience the capability for analysis, they become interested in using multiple dates of data on a continuing or operational basis. The advantages include: examination of wetland quantities under different hydrological or meteorological conditions; use of multiple dates to facilitate identification of marginal or temporal wetlands; increasing sample size to help fix the position of wetland boundaries; avoiding errors of commission or omission due to use of limited sample size; avoidance of errors due to unusual tidal conditions; and other reasons.

The potential disadvantages are: commitment of personnel, resources, and funds; the planning for long-term use of the GIS database to amortize costs over years; and/or limiting the cost associated with one-time use of the database.

A lot of multiple date inventories and analyses of wetlands using GIS have been completed in marine and freshwater systems. Some of the best evaluations have been made on the Great Lakes. This is due to the fluctuating quantities and varieties of coastal and inland wetlands resulting from the month-to-month and year-to-year variability in Great Lake water levels. Unlike marine systems, tidal influences are very small on the Great Lakes, and studies there provide good examples for applications of GIS to other lakes and reservoirs. These applications are supplied later in the text.

## Features of Interest

Much of the analysis of remote sensor data revolves around the presence/absence of features (Figure 4.5). This is due here to the importance of features for optimal identification and later monitoring. The features of interest include: topography, soils, hydrology, and land ownership. Other features of interest are those that indicate favorable conditions for preservation of wetlands, construction of wetlands, and supply valuable data for maintenance and monitoring of wetlands and related ecosystems.

On an operational basis, it is desirable to establish periodically the characteristics of wetlands. This would involve a periodic analysis of local conditions, such as type, extent, functions, value, and other conditions. A program of monitoring would be appropriate to address these interests.

Adjacent locations of features are important data in understanding and managing wetland properties. Basic data requirements could also include data on the surrounding characteristics of a given site. Use of remote sensor imaging data and a GIS system would allow monitoring data on adjacent resources or land cover to be incorporated, and can be used along with data on the wetlands themselves. Appropriate data includes: location of the public and their activities; location of built or constructed features; other features; and location of features with respect to wetlands and facilities.

*Figure 4.5.* Wetlands and unconsolidated shore areas can be habitat for endangered species and require special management, such as the Straits of Mackinac, Michigan, where the piping plover nests.

A number of characteristics and conditions should be evaluated. It is necessary to know the exact location of the wetlands. These characteristics can be derived from maps, field visits, and surveyed characteristics.

Periodic monitoring with aerial photographic, aerial videographic, and/or aerial or satellite remote sensor technologies can supply these general data. Remote monitoring can also be accomplished at lower relative costs. Various imaging technologies have certain capabilities and costs. Orthophotography, for example, is very capable of producing large-scale map sheets. However, it can be relatively expensive for certain applications. This is as compared to other, traditional data acquisition methods such as field sampling and field mapping.

*Figure 4.6.* Development often alters the drainage conditions of areas, and often retention ponds are dug to handle storm runoff. Some wetland habitat is created but it many lack many wetland functions such as this pond near Syracuse, New York.

## Information for Planning

The benefits to wetlands of a remote measurement and GIS system are multifold. Of particular value are remote imagery data and spatial data systems in the planning phase of a project (Figure 4.6). For planning, the data are useful in optimization of avoidance of wetlands, optimization of management activities and relative economics, and other efforts.

The use of remote sensor data and spatial information systems like GIS can be valuable also in reporting information. The planning phase of a project or ongoing management will often require information for permitting and zoning, and general reporting to the government, the public, and to management personnel.

A recent concern is the extent and variety of wetlands on the continental and global scale, as researchers and managers become concerned with global change issues such as the source of radiation-important trace gases including methane and nitrogen. To address wetland resources at either the local or global scale requires identification, inventory, and monitoring to supply information for informed decisions.

The personnel that deal with these problems and questions will be best served with a computer, imaging, and spatial analysis capability. This will meet the needs of agencies or companies for operations and management (Figure 4.7), or planning information. It will also potentially provide for any governmental, conservation and environmental group, industry and/or industry-government needs, or requirements for sharing of data.

## Indicators of Water

The movement of water can cause a number of problems for management of wetlands. In the past, many wetlands have witnessed both unusual disasters and unusual meteorological events which act as natural stressors and challenge the integrity of wetlands (Figure 4.8).

It is important to monitor wetlands for usual wetland characteristics and conditions. It is also important to have the capability to monitor following unusual events or even a disaster.

*Figure 4.7. Wetlands often develop in drainage ways, but these are often mowed to increase water flow and visibility for drivers, as in Seneca County, Ohio.*

Remote measurements of water characteristics and conditions can be particularly helpful. Water appears much differently on imagery. During and following events, standing water and wet areas can be identified by the presence and abundance of water. These characteristics can assist in locating the results of flooding such as inundated areas and areas for potential risk.

Distribution and duration of flooding can be identified through use of operational remote sensing systems. The use of visible, near and thermal infrared and radar sensors and their measurements can supply the required data.

In particular, the advent of satellite radar sensors offers great capabilities in detection and inventory of wetlands (Lyon and McCarthy, 1981; Ramsey, 1998), and monitoring of flood activities. Satellite radars operate at wavelengths that pass through cloud cover. These radars make their own microwave light and thus are imaging the earth day and night with active sensing (Lunetta and Elvidge, 1998).

*Figure 4.8. Drainage of storm runoff is vital to communities and maintenance is necessary. These conveyances are often high quality habitat in urban areas such as Henderson, Nevada.*

## General Condition of River and Stream Wetlands

The remote sensing approach can allow for certain specific evaluations of high-valued riverine wetlands. Potentially, remote sensor measurements can help to identify areas with undermining of river and stream crossings. Remote sensors can sense the extent of flooding adjacent to a bridge crossing, for example. After the flood, remote sensors can identify the extent of deposition and erosion of floodplain sediment materials (Figure 4.9). These indicators can help identify areas for checking by field personnel, and potential avoidance of the failure of structures and resulting exposure to wetlands and riparian areas.

## Other Risks and Impacts

There are a number of activities or events that can cause risk to wetlands. Many of these activities or events can be detected. Many can be identified from indicators or surrogate variables, and often early identification can lead to remedial actions.

In certain areas, the conditions are much different from other areas and they present problems unique to the location. Often these conditions are the result of a combination of geological or hydrological characteristics.

*Figure 4.9. Desert areas such as Tucson, Arizona often experience high intensity rain events and flooding which may produce ephemeral wetlands and wetland functions for short periods of time. Yet this short period may be very important in arid or semiarid areas.*

Many areas have long-term or historical problems with erosion or mass wasting. These phenomena contribute unwanted sediment and rock to wetland resources. For example, volcanic areas have specific problems associated with the geology. These areas of the U.S. and Canada include: Alaska, Hawaii, British Columbia, Alberta, Washington, Oregon, California, Idaho, and others. The potential impacts in active volcano areas include: burial; explosions; fire; earthquakes; tsunamis; and other impacts. In historically inactive volcanic areas potential impacts include: landslides; rockfall; earthquakes; and other impacts caused by erosion and/or mass wasting of the silty-clayey soils derived from most volcanic parent rock materials.

A major interest is the identification and location of human activities that constitute environment risk near or in wetland ecosystems. Due to regulatory considerations, the human activities or features in or adjacent to properties need to be identified and monitored.

Monitoring of activities can be accomplished through the use of aerial photography and remote sensor measurements (Figure 4.10). The identification of an activity involves change in land cover or water resources. Definitions of land cover include the characteristics of the land surface as to type; e.g., forest, agriculture, urban, suburban, and other. Changes found in water resources can include increased sedimentation, algal blooms, loss of riparian vegetation, and the like.

*Figure 4.10. Older urban and industrial areas often have wetlands and they supply wetland function in areas where few local wetlands persist, such as in urban Marquette, Michigan.*

## Identification of Habitat for Conservation

Identification of habitat for species has been a consideration in siting a constructed wetland or selecting property for acquisition for mitigation of wetlands. Much like any development, a checklist of items is addressed in selection and acquisition of a wetland. New items for the checklist include: presence/absence of plant and animal species of interest; presence/absence of general or jurisdictional wetlands; presence/absence of cultural resources; and other features.

One GIS method to address plant and animal species is to take advantage of governmental data sets. This also serves to leverage industry work. Generally, data on the distribution of plant and animal species are hard to find. Seldom are data in digital form, but they exist as records and/or maps.

A program of the U.S. Fish and Wildlife Service (USFWS) and now the USGS is creating data sets of plant and animal distributions and habitats. The GAP program is developing on a state-by-state basis a GIS database on species. For each plant or animal of interest, a GIS file or layer is composed of distribution and habitat information. These data will be or are available state-by-state, and can be incorporated into a GIS system as part of the data inputs.

Management of diverse data types such as GAP data can be accomplished using GIS. It provides for a lot of input data types, and will help to leverage on-site work by avoiding the cost of generating these data independently.

*Figure 4.11. Where wetlands are removed for urban expansion they are often mitigated on-site. Here a deep pond has been partial filled and wetland plants established. Unfortunately, lack of continued maintenance beyond the permit requirement has resulted in their disappearance from central Ohio.*

## Human-Related Features and Wetland Mitigation

There are a number of features of wetlands that may be influenced or related to humans. Regulatory agencies and the public now require additional examination of human-related issues, and the presence or absence of these features can be important in selecting lands for construction of wetlands for habitat enhancements or mitigation (Figure 4.11).

Cultural resources can greatly impact engineering design, permitting, and construction of mitigation wetlands. Frequently, these resources and potential influences of construction and operations of wetlands come into question.

Ancient sites may be encountered in a number of locations, either rural or in urban and suburban areas. Identification of probable locations is important, and will potentially influence locations of wetlands and their maintenance.

Restricted properties may also influence the selection of land appropriate for use as a constructed wetland. These conditions are known to industry, and available records or databases can be consulted. Sources can include state and local historical societies, and anthropological/archaeological groups. The data derived from records and maps can be incorporated in the GIS to be used in selection.

*Figure 4.12. States with deserts are concerned with identifying and maintaining ephemeral wetlands. Often these can be found along intermittent streams and riverine such as this example near Lake Mead in Nevada.*

Incompatible land ownership or land not available for easements or for sale presents a real problem (Figure 4.12). Prevailing conditions can be noted and stored for analyses. Schools, cemeteries, restricted federal and state ownership, and other ownership may be important. Future considerations may include criterion such as house counts in a given area or population.

## Hazardous Waste

Identification and avoidance of waste sites is important to selection of property to create new wetlands. Identification and avoidance of hazardous wastes and buried tanks can be facilitated by examination of data sources including maps, records, and images. Historical aerial photographs are particularly useful for identification of change in land cover and types. It is possible to identify characteristics and conditions of features that indicate the potential of these activities on a given site (Lyon, 1987). It is a valuable source of data and an adjunct to site visits and other evaluations.

# 5 *Imagery and Interpretations*

A goal is to provide better information on wetland ecosystems as they are influenced or exposed to ecological and environmental risks. It is desirable to utilize aerial photo and remote sensor data in combination with Geographic Information Systems (GIS) technologies to address these ecosystems and facilitate analyses. In particular, people wish to work on: (a) better methods of collecting field data with in situ and remote sensor measurements; and (b) integration of sensor data of varying resolutions for input into Geographic Information Systems.

While the potential for remote sensor and other point or area sampling data is great, it remains to further create operational applications of these data in developing GIS activities. To date the limited, operational use has resulted from the difficulty of mastering technologies used in GIS and remote sensing, and some technology limitations (Johnson and Handley, 1990; Deysher et al., 1995; Elevand et al., 1995). A combination of techniques is capable of addressing the problems above and assisting in the measurement and modeling of resources for management purposes.

Monitoring of wetland ecosystems and the exposure of stressors on wetlands or adjacent ecosystems can best be met by an analysis of multiple sources of remote and in situ sensor data, GIS databases, and models of wetland and water resource characteristics (Field et al., 1990; Jensen et al., 1992; Lunetta and Balogh, 1999). Monitoring experiments require good quality data for initialization of the system and real-time delivery of data. It remains to address several areas of research before operational systems can play a major part in ecosystem or risk evaluations. These areas include: (a) scaling of data and compiling databases of different raster, sensor data for use by GIS and other, traditional models; and (b) integration of in situ sensor and other point sampling data in models, and GIS databases for near real-time evaluations.

# Remote Sensing

A valuable tool in the identification and characterization of these wetland ecosystems and related land cover types are remote sensor technologies. These technologies measure and store the characteristics of variables of interest or related surrogate variables in a permanent record. They allow the collection of data over large areas in a relative short period as compared to a 100% field sampling of the ecosystem.

Remote sensor technologies may also operate in parts of the electromagnetic spectrum that go unmeasured by the human eye. Often the wetland and land cover types of interest behave differently with respect to the reflected or upwelling light as compared to terrestrial land covers (Figure 5.1). This capability to measure and analyze the differential light reflectance characteristics of materials can be exploited to help identify the ecosystems of interest.

*Figure 5.1.* Low altitude aerial photographs can be acquired to record the characteristics of the wetlands and surrounding areas. Riverine wetlands and riparian areas can be identified by the crowns of trees and shadows. A small pond is apparent with emergent wetlands. At the top center is a cornfield with a gully highlighted by the shadows.

Remote sensor technologies run in complexity from simple camera and film systems that measure broad parts or bandwidth to advanced hyperspectral sensors that measure upwelling light in very narrow parts or bandwidth parts of the electromagnetic spectrum (Mynar and Lunetta, 1990). The level of sophistication of tool can be selected as compared to the requirements of the application. Due to the variety of remote sensor technologies, there is usually an example technology that will meet the requirements of the application, and do so at a cost that is reasonable as compared to the project scope and budget (Falkner, 1994; Lyon et al., 1995).

Remote sensor data are acquired in digital format or as analogue film products (Figure 5.2). Each product type can be handled digitally to facilitate computer processing. The digital format and spatial and temporal characteristics of the data make them valuable for GIS applications. It is no mystery that many GIS applications utilize remote sensor data in many applications.

## Photointerpretation

Remote sensing technologies as defined here include a variety of tools ranging in complexity from advanced sensors to simple camera and film systems. Each hardware system creates two-dimensional image products that can be analyzed or interpreted for valuable information. The process of analyzing these images is called interpretation.

*Figure 5.2.* This low altitude photo shows how stream courses are sometimes devoid of shrubs and trees. In the center is a wooded area bordering the stream, which might be a wetland area.

The generally accepted definition of image or photointerpretation incorporates several steps or elements, including the fact that interpretation is an art and science of obtaining or interpreting data from the characteristics of features recorded on photographs or images. Photointerpretation refers specifically to interpretation of photographic products, while image interpretation is more general, as it covers the interpretation of all image products including photographic examples.

Generally, these terms refer to interpreting images or photographs taken from aircraft or spacecraft. The methods used for interpretation are the same methods and skills that can be used to obtain information from any type of photograph or image.

Interpretation applies several methods through skillful application to obtain details about features found in images. These elements of interpretation supply information on features that are basically independent assessments of each characteristic. These elements include characteristics such as: color or grey tones, shape, size, texture, pattern, shadow, and associations (Figure 5.3).

The grey tonal range supplies detail about features or materials in a range of contrast between black and white. Typical materials exhibit a range of grey levels on visible black-and-white photographs. Rock, bare soil, concrete and the like often appear light-toned. Vegetation is relatively dark-toned due to the low relative reflectance of green plant material compared to bare soil or rock.

The variability of visible light can be measured through the use of color film. Colors show the tonal variation of each additive color primary, such as ranges of blue, green, and red colors. Color, like tone, can be used to evaluate features as well as pinpoint the spectral variability of materials in the visible portion of the spectrum.

*Figure 5.3.* Wetland areas are often created by impoundment. The earthen dam is identified by the straight line of the water body. Wetlands are found where the streams enter the impoundment.

Shape refers to the exterior configuration of materials or features. Many features or objects have distinct regular or irregular shapes. Shape can be a unique clue as to the identity of the feature.

Size refers to the absolute or relative dimensions of the object or feature (Figure 5.4). Size can be very important as features are often indistinct on photographs and knowledge of size can be a big clue as to the identity of a given feature.

Texture is a variation in tone or color caused by a mixture of materials on a given site. Interpreters will identify a texture as a faint variation in tone or color. In an abandoned farm field the texture may result from a tonal or color contribution from the presence of shrubs or small trees in the field. In the eastern U.S., the trees may be sumac or aspen, and in the western or mountainous U.S. they may be small conifer trees or shrubs. These shrubs or trees are too small to recognize by their individual crown or canopy vegetation. Yet they contribute to the overall variability of tone as compared to the relatively uniform tone of a mature soybean or wheat

*Figure 5.4. Aerial photographs are particularly valuable where access is difficult and wetlands are large in size. Pictured are riverine wetlands along the Mississippi north of La Crosse, Wisconsin and south of Minona, Minnesota.*

crop in a field. The presence of tall plants, partial canopy closure, and shadow help create a textural difference that often allows the separation of corn crops from soybean crops.

Pattern, conversely, is the regular or irregular distribution of relatively large features on the earth's surface. An example is the regular pattern of field boundaries in a rural area. When this pattern is disturbed, say by a stream course, one can infer certain characteristics of the stream channel such that it is too big to be altered by human activities over time (Figure 5.5). Often, these channels will be obscured by vegetation such that the water itself is hard to see. The pattern of the riparian vegetation still indicates the true characteristics of the feature and the presence of vegetation that shades the stream and offers habitat for biota.

The shadow of a feature can be an important clue and may help identify the feature as to type. The shadows cast by a feature can provide additional important shape and size information. A common example is that of interpreting the name of a business or building from the shadows of individual letters that form a sign. The individual letters "blend" into the building and/or may be too narrow to view from above, but the shadows are distinct to the interpreter.

Conversely, shadows can obscure detail. It is difficult to view a feature hidden by a shadow, due to the lower relative illumination in the shadow area as compared to the overall illumination of the sunlit areas.

The "association" of a feature is the other characteristic or clue that is found together, in association, with the feature of interest. For example, the course of a stream is usually evident by shape, but often the tone of the water is hidden from view by trees. Stream courses are usually found in association with other clues such as: the meandering stream pattern; the branching drainage pattern; automobile and railroad bridges; pond or lakes; streamside vegetation; and lower relative elevation and a downhill course as compared to the surrounding landscape.

## Interpretations

Interpretations of images can supply a great level of detail on wetlands, water features, and associated land cover types. Use of the elements of image interpretation along with knowledge of the features of interest and experience at interpretations can provide valuable information.

*Figure 5.5. Photographs can be acquired whenever the weather is suitable for flying. Sometimes, winter scenes can display a lot of detail on wetlands because snow cover often is different as compared to upland and river areas in Missouri or Iowa.*

When viewing images, interpretations supply information on wetlands and aquatic ecosystems (Niedzwiedz and Batie, 1984; Niedzwiedz and Ganske, 1991). One can see details of the nearshore topography or bathymetry due to the penetration of visible light into and back from the water. Often, the shape of submerged, coastal aquatic features can be identified due to the penetration of visible light into the shallow water column. One should note the very light tone or color of bottom sediments, roads, and beach areas (Figure 5.6).

In comparison, the vegetation of the submerged aquatic plants or submergent wetlands can be identified in shallow water areas due to the dark relative tone or color as compared to the surrounding light-toned sand or sediments. The contrast between light-toned

*Figure 5.6.* Aerial photographs are acquired by the government on a regular basis and are available from archives. Here the presence of riverine systems of the Columbia and Willamette have formed deposits which were developed into urban and industrial areas, including a port. Note the presence of lacustrine and riverine wetlands near Portland, Oregon.

or colored bottom sediments and the relatively dark tone or color of submerged vegetation allows for the identification of features of interest.

Such interpretation clues can be articulated for a given ecosystem under study, and formalized into a list of characteristics or interpretation key to facilitate identification and inventory. This approach has worked well in a number of studies, and is used to great effect in characterizing submerged aquatic vegetation (SAV) that is the focus of habitat studies for a number of species (Raabe and Stumpf, 1995).

The use of differences in spectral reflectance, as described above, can facilitate a number of applications. Analysis of reflectance characteristics shows that dark-toned features found beneath or above the water are emergent and submergent vegetation. They appear dark because plants reflect little light in the blue part of the visible spectrum and in the red part of the spectrum. On a color or black-and-white image these plants will be dark tones or dark color, as the green reflectance of light by plants is relatively low compared to that of soil.

Interpretations are facilitated by the evaluations of differential spectral reflectance of a variety of portions of the electromagnetic spectrum. The use of a number of images and other different film types can be a great help in interpretation (Lyon, 1987; Williams and Lyon, 1991; Lyon and Greene, 1992; Lyon, 1993; Lyon and McCarthy, 1995). This is particularly true of hydrological characteristics of areas which change over short or long periods of time. Multiple dates of photos or images and multiple types of films or sensor band data can help identify the general hydrological characteristics of wetland areas by capturing wet, dry, and normal moisture periods.

Following rainstorms and other hydrological-related events the water may be opaque due to the runoff of and resuspension of sediments. These suspended sediments or nonpoint pollutants may be difficult to image through, and can disrupt evaluations of submergent wetlands (Lyon et al., 1988). Conversely, the suspended sediments may help to locate the movement of water and waterborne pollutants. Multiple images can help lend insight into a variety of conditions by capturing the variability over time.

It is always useful to obtain multiple dates of aerial photos or remote sensor data of a given study area in support of analyses. This is because each photo supplies unique information, and repetitive coverage adds to the value of multiple samples and statistical analysis capabilities to the project. Photos may also be inexpensive in comparison to the costs of obtaining the similar quality and quantity of field data using traditional methods.

Photos and images are also valuable in locating the position of features and materials. Knowledge of size and position can be supplied by the measurement capabilities of images (Figure 5.7). Analysis of images can provide horizontal and vertical information by using a combination of photogrammetry or surveying technologies. Photogrammetric measurements can be made on the images and tied to some absolute reference to characterize the location of earth or terrestrial features in an absolute sense. Surveying also allows one to later relocate these features using the original measurements or map products.

Imagery and Interpretations

*Figure 5.7.* Historical aerial photographs allow the viewing of riverine systems, including the Niagara River of New York and Ontario.

## Historical Aerial Photographs

Another valuable source of data in planning and monitoring is historical aerial photographs. Aerial photographic coverage of the United States (U.S.) is available from archives. Generally, photographs date to the late 1930s. Multiple sets of aerial photos are in archives, and one may be able to develop a time series of photos extending from the 1930s to the present (Figures 5.8, 5.9, 5.10).

*Figure 5.8. Historical photographs are very useful in the study of wetlands because they capture the resource at various times and under various hydrological and meteorological conditions. Note how sun glint or specular reflection identifies some water areas near Sauvie Island, Oregon and Washington, in the bottom center (July 5, 1970). North is to the right.*

The interpretation of historical aerial photos provides data on a number of conditions. From individual dates of coverage, interpretation will yield data on the land cover types, presence or absence of houses and buildings, stream drainage pattern, and general soils and geomorphology characteristics. Multiple dates of coverage allow the user to capture the different hydrologic conditions of wetlands, lakes and coastal areas that have occurred over time (Carter, 1982; Butera, 1983; Lyon, Drobney, and Olson, 1986; U.S. Army, 1987; FICWD, 1989; Williams and Lyon, 1991; USEPA, 1991; Lyon and Greene, 1992; Lyon, 1993).

Sources of historical aerial photographs at the federal level include: the National Archives and Records Service in Washington, D.C. for pre-World War II photographs; the

*Figure 5.9. Photographs taken at a different time yield some of the same information and supply other details that are new, such as increased data on tree crowns found in the forested wetlands on Sauvie Island.*

U.S. Geological Survey EROS Data Center in Sioux Falls, SD for U.S. Department of Interior agency photographs; and the Aerial Photography Field Office in Salt Lake City, UT for U.S. Department of Agriculture (USDA) agency photographs. The addresses for these sources are available from the Internet or in Lyon (1993), Lyon and McCarthy (1995), and Ward and Elliot (1995), and from other sources.

Aerial photographs or remote sensor images are usually available for every other year or every third year in the USGS and USDA archives. It is possible to gather approximately 10 or more dates of aerial coverage since 1930s from the sources above (Lyon, 1981; Lyon and Drobney, 1984).

Another good source of aerial photographs is the U.S. Department of Agriculture Farm Service Agency (FSA) or former Agricultural Stabilization and Conservation Service (ASCS). Since approximately 1981 the ASCS, now FSA, has collected small format (35 mm) color transparency aerial photographs of farmed areas subject to crop support programs (Lyon,

## Wetland Landscape Characterization

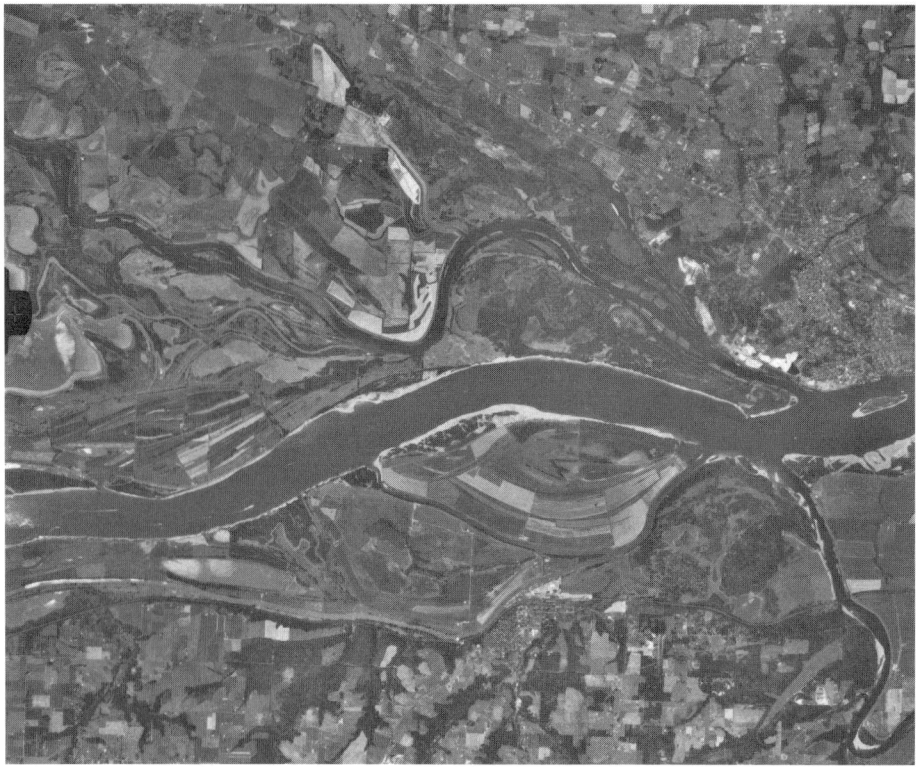

*Figure 5.10. Different governmental groups acquire photographs for different reasons, and hence there are a variety of scales and film emulsion types to be selected from archives (1975).*

McCarthy, and Heinen, 1986). The photos were taken on color slide film and acquired at relatively low altitude, creating large-scale photos for analysis. These photographs have been archived within each state since approximately 1983, and recent coverage (through approximately 1995) is archived in the county or state FSA offices where they can be viewed and ordered.

Significantly, these photos have been acquired during the growing season of crops and other vegetation. Regular aerial mapping photographs are usually acquired during the dormant or leaves-off period of vegetation. Leaves-off photographs are valuable so that the stereoplotter operator can follow terrain contours and make topographic maps without the land surface being obscured by leafy vegetation (USACE, 1993; Falkner, 1994). Because the FSA photographs were acquired in the growing season, the wetland and aquatic vegetation was in active growth, as compared to the spring or fall period of leaf-off conditions that are usually specified for the acquisition of mapping photographs.

It is still possible to identify and inventory general wetland areas using leaf-off photographs. This is because the extent of the wetland area is defined by the vegetation residue present from previous growth, and from the difference in wetland appearance as compared to adjacent terrestrial or aquatic systems (Lyon, 1993).

## Characterization of Vegetation

An advantage of using photographs or sensor data from the near infrared part of the spectrum (0.7 to 1.1 µm) is that healthy, vigorously growing plants reflect a great deal of near infrared light. These plants or vegetation appear bright, white-toned on black-and-white infrared photos, or bright pink or magenta on color infrared film or sensors. These characteristics allow the user to identify growing or green plants from dead plants or plant residue. The user can identify wetland plants growing above the surface of the water, or emergent wetland plants. This can be accomplished by using near infrared products and by identifying the bright, white tone on black-and-white infrared photos or bright pink color on color infrared film or sensor data.

Conversely, plants growing beneath the water surface, or submergent wetland plants or submergent aquatic plants, do not exhibit the bright tone or pink color of emergent plants because the near infrared light is largely absorbed by water. This difference allows the user to identify emergent wetland plants and terrestrial plants, and distinguish them from submergent wetland or submergent aquatic plants which appear dark-toned or dark blue in color on black-and-white or color infrared film or sensor products (Lyon and Olson, 1983; Lyon, 1993).

As addressed in previous sections, the appropriate methods for characterizing vegetation in wetlands are based on the project goals, and the scale or resolution of the application. Depending on scale, vegetation communities can be identified by plant species in the field. The characteristics of the vegetation communities of plant species can be identified from a distance by their shape, size, tone or color, and other diagnostic characteristics. Remote sensor data can be particularly helpful at identifying vegetation communities at various scales.

## Soil Characterizations

Photointerpretation of aerial photos and remote sensing experiments have been performed over the years with an eye toward characterizing soil features and moisture conditions. Recently, great work has been done using the middle infrared portion of the spectrum, because it is sensitive to moisture conditions of soils and plants. Good work has been done using different dates of Landsat TM data and the middle infrared bands 5 and 7 to identify hydric soil conditions. Lunetta and Balogh (1999) found that the combination of spring, leaves-off imagery and growing season imagery allowed for increased detection of hydric soil conditions that are often obscured by leaf canopy cover in the growing season.

Remote sensor data can also be used to help distinguish between mineral soils and soil

*Figure 5.11. Very high altitude photographs are acquired by NASA such as this scene of the southern San Francisco Bay area of California. The presence of a variety of wetlands can be detected and measured including tidal wetlands, diked wetlands, filled wetlands, and diked areas for the distillation of water to form salt deposits.*

high in organic matter. The former will have a different color or spectra in the red and green part of the spectrum, as compared to the dark tone or spectra of the very organic soils.

## Water Resource Characterization

Remote sensor data and field data are particularly good for identifying hydrologic characteristics of wetlands over time (Figure 5.11). Remote sensor data records the reflected or upwelling light from water and the result is good detail on water or aquatic ecosystems as distinct from terrestrial ecosystems. The near infrared portion of the spectrum has been used to identify plant and hydrologic conditions of wetlands. This is due to the unique light or spectral characteristics of water and plants on color or black-and- white infrared film, or from instruments measuring light in the near infrared portion of the spectrum.

The majority of near infrared light is absorbed by water. A very small portion is reflected or transmitted through water. The net result is that water with very low concentrations of suspended sediments or chlorophyll in phytoplankton will absorb most infrared light. This causes water to appear dark-toned or dark black-blue on black-and-white and color infrared films, respectively (Lyon and Olson, 1983; Lyon, 1993; Ward and Elliot, 1995).

The behavior of near infrared light and water facilitates the interpretation of certain conditions of wetlands, lakes, and coastal areas. The first advantage of near infrared film or near infrared sensor instrument measurements is the detection of the edge of water and soil. Near infrared light reveals the edge between water and shore because very small thicknesses of water will absorb near infrared light, and the edge is identified by a dark tone or dark blue color on black-and-white and color infrared film, respectively. Hence, infrared film or sensors can identify the edge of water bodies.

Conversely, film or sensors working in the visible part of the spectrum experience penetration of light through the water to the bottom and reflectance to the surface (Lyon et al., 1992; Lyon and Hutchinson, 1995).

## Wetland Classification

In inventory and mapping of wetlands and other land cover types, it is necessary to identify the wetlands as to category or type. Often this activity is called thematic mapping of land cover "themes." This is similar to the mapping of geological types or themes or mapping of land ownership.

There are a number of types of wetlands and classification systems to describe them (Anderson et al., 1976; Cowardin et al., 1979; USACE, 1987; FICWD, 1989; USEPA, 1991; Brinson, 1993). These types and classification systems have something to do with the hydrological, soil, or edaphic and vegetation characteristics of wetlands. Other systems that characterize wetlands have something to do with the functions and uses of wetlands (Adamus et al., 1991; Brinson, 1993). Still, other description systems have something to do with the scale of the identification or inventory of wetlands.

A commonly used system in the United States is the Cowardin System of the U.S. Fish and Wildlife Service (Cowardin et al., 1979). It is employed in describing wetland areas as to types, and for mapping wetland types in the National Wetland Inventory (NWI) and their series of map products.

## Creation of Aerial Photo Land Cover Products

Land cover classifications and map products can be made according to the project goals or scope using a team of photo interpreters led by an expert. The team can use the mentioned data sources to map land cover polygons according to a scope, and conduct the interpretation and classification process to produce the desirable land cover types.

Polygons can be interpreted and marked on transparent overlays of high altitude photographs. Polygons should meet the minimum mapping unit requirements described in a scope, and be drawn and identified using the land cover classification types.

A member of the interpretation team should work on the individual quadrangle areas and photos. The overlays can be made of polygons and of their land cover type. Work should be tracked by photo interpretation worksheets containing information in numerical and in written form (Lunetta et al., 1993).

Particular attention should be taken to assure correct land cover classifications. This includes training and monitoring the work of the photointerpretation team by the expert (Congalton, 1996; Congalton and Green, 1998).

Particular attention should be paid to identifying wetlands, using guidance from a scope and using the techniques described by Lyon in *Practical Handbook for Wetland Identification and Delineation* (1993), which provide a number of valuable methods for inventory and classification of wetlands. Included are descriptions of methods to conduct a large-area or regional evaluation of wetland land cover types.

The quality assurance and quality control plan should be in place to track the photointerpretations by photo and quadrangle using the worksheets. Each overlay of a given photo and its worksheet should be evaluated for the positional quality and classification accuracy by the expert. In this manner overall quality is evaluated by the expert, and any problems are identified quickly. Solutions to problems should be recorded and distributed to the photo interpreter for discussion and action (Congalton, 1991).

The resulting transparent overlays can then be transferred photogrammetrically to the USGS quadrangles and digitized.

## Detection of Change Methods

A question that is commonly asked about wetlands is, "How much do we have?" or "How much have we lost?" For comparison purposes, the following questions are, "How much was present historically?" and "How much will we have in the future?" The public poses these questions all the time, and they are the main questions asked by decision-makers. Hence, it is vital to obtain data that can establish historical conditions and detect change over time, to establish trends.

It is often difficult to develop data for the historical extent of wetlands (Figures 5.12, 5.13), or for that matter determine the historic extent of ephemeral lakes, or of the historic

*Figure 5.12. Comparison of more recent photographs to historical examples allows one to identify the wetlands and any changes in land cover over time. Pictured here is the Sleeping Bear Dunes National Lakeshore and Lake Michigan (1975). North is to the left.*

position of the coastal zone. There are several methods to develop historical numbers of general wetlands. The sources of information can include old maps, survey records, and governmental records. Naturally, these are subject to a certain potential for errors, either known or unknown. It is also sometimes difficult to reconcile the definition of a given resource in history, as compared to the definition used in the current time frame as a goal for the project.

Lake and coastal wetland ecosystems exhibit changing characteristics due to the variability of water resources. The lake system experiences influx of water from runoff, ground water and interception of precipitation. The origin of the lake and the characteristics of the watershed largely govern lake water quantity and quality issues, and govern to a certain extent the quantity and variety of wetlands. Likewise, coastal wetland areas are defined by water resource characteristics and range from freshwater coastal lakes to marine coasts to the coasts of hyper-saline lakes or seas.

The methods used to evaluate wetland characteristics and extent, and supply data for GIS analyses are often set by the goals of the project and by the scale of the effort. Lakes are

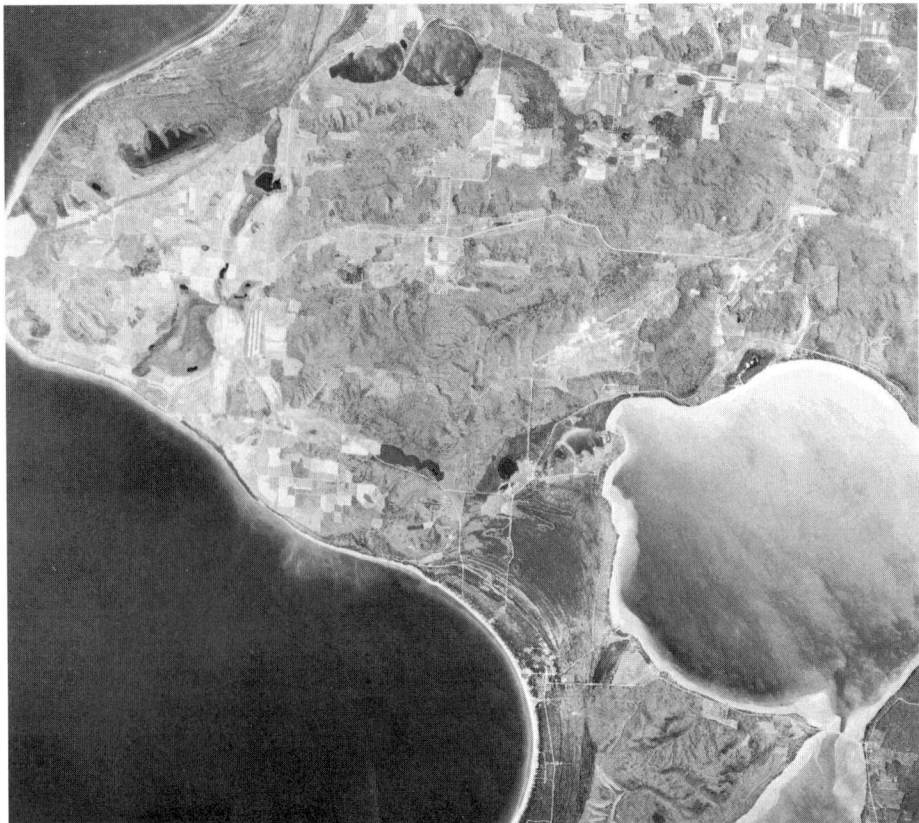

*Figure 5.13.* Historical photographs reveal the earlier conditions of the land as well as water resource characteristics including shallow areas of Glen Lake (1953).

also greatly affected by the three-dimensional influence of water depth characteristics, and the relation between water depth and forcing functions or stressors such as wind-driven currents, thermal stratification, and evaporation and transpiration issues in shallow lakes.

Coastal wetland areas exhibit some unifying characteristics and some greatly differing characteristics. By definition coastal areas are shallow in water depth as compared to the depth of the open lake, sea, or ocean. Many of the techniques used to characterize coastal areas are the same as those used in lake and wetland areas (Thompson and Gauthier, 1990).

Many evaluations of wetlands and their change over time have been completed using aerial photographic products and interpretations (Scieszka, 1990; Thompson and Gauthier,1990; Sherin and Edwardson, 1995). Evaluations of multiple dates of historical photographs allow interpretation of conditions and changes over time. It is recommended that one use four or more dates of photographs, if possible, to look at various features over time (Lyon, 1987; Lyon, 1993).

From a number of dates of historic photographs, hydrological, vegetation, soils and

*Figure 5.14. The advent of space radar satellite sensor allows all-weather sensing, and details that may not be supplied by visible sensors. Note how the topographic characteristics are more evident in this SEASAT image from 1978.*

human activity data can be interpreted. It is possible to evaluate the presence/absence of hydrological conditions or stressors and their exposure to general wetland areas. These efforts have been described here and elsewhere.

Many current efforts have made use of digital satellite remote sensor products for analyses of change (Arnold, 1992; Balogh et al., 1992). The future holds many opportunities for these sort of digital remote sensor analysis efforts, as this approach allows for more specific identification of wetland types or receptors based on spectral characteristics in the visible, near, and middle infrared, and microwave portions (Figure 5.14) of the electromagnetic spectrum (Lyon and McCarthy, 1995).

Another advantage of remote sensor methods is the utility of computer processing of data, and storage and modeling of data using Geographic Information Systems (Sinclair, 1990; Young and Dahl, 1995). The advent of computer processing has added great analytical power to evaluations, including detection of change in wetlands (Figures 5.15, 5.16).

## Wetland Landscape Characterization

*Figure 5.15. Satellite images are often available in pictorial form and in digital form. Here is a series of images of the Columbia River and surrounding ecosystems from the Landsat Multispectral Scanner (MSS) sensor operating in the near-infrared. Note how dark water features appear such as the Pacific Ocean and Columbia River (July 29, 1972).*

## Detection of Change Using Satellite Data

Change detection methods have been developed over time, and generally take the form of methods that work on data in both pre- and postcategorization image products (Singh, 1989; Lyon et al., 1998). For the precategorization change detection techniques, only those areas that have undergone significant spectral change will be categorized.

Efforts have been made over the years to develop methods to efficiently and accurately evaluate land cover change detection from satellite remote sensor data (Colwell and Weber, 1981; Pulich and Hinson, 1995; Elzinga and Evenden, 1997). These procedures were initially developed on ecosystems in North and South America including the trans-Amazon, for purposes of evaluating global change (Hastings and Di, 1994a, 1994b). Due to the importance of change detection in research and management missions, large efforts have been devoted to development of these procedures for operational evaluations of change from AVHRR,

*Figure 5.16. The same location and time as in Figure 5.15 is displayed for the red portion of the electromagnetic spectrum. Note the dark tone of forested areas and light tone of agricultural and urban areas. Water can range from dark grey to almost white, based on surface-suspended sediment concentrations (July 19, 1972). Sauvie Island is in the center of the image.*

MSS, and TM data sets in a variety of studies (Lee and Lunetta, 1995; Leshkevich et al., 1995; Ramsey et al., 1998).

Other projects have focused on global change issues related to forested areas in the Pacific Northwest and in the State of Chiapas, Mexico, as part of the North American Landscape Characterization (NALC) project (Lunetta et al., 1993; Lyon et al., 1998; Lunetta et al., 1998). An initial NALC pilot project to test the land cover classification and change detection procedures was conducted within the 64,000 square mile Chesapeake Bay watershed (Lunetta et al., 1998).

Early efforts at using Landsat MSS data for the detection of land cover change were largely based on visual interpretation of the multitemporal images. This was essentially a photointerpretation process, using techniques developed in the decades preceding the launch of Landsat in 1972. This method is still widely used for change detection (Figures 5.17, 5.18). It is perhaps the most direct and fastest change detection technique, particularly in the initial stages of a change detection project.

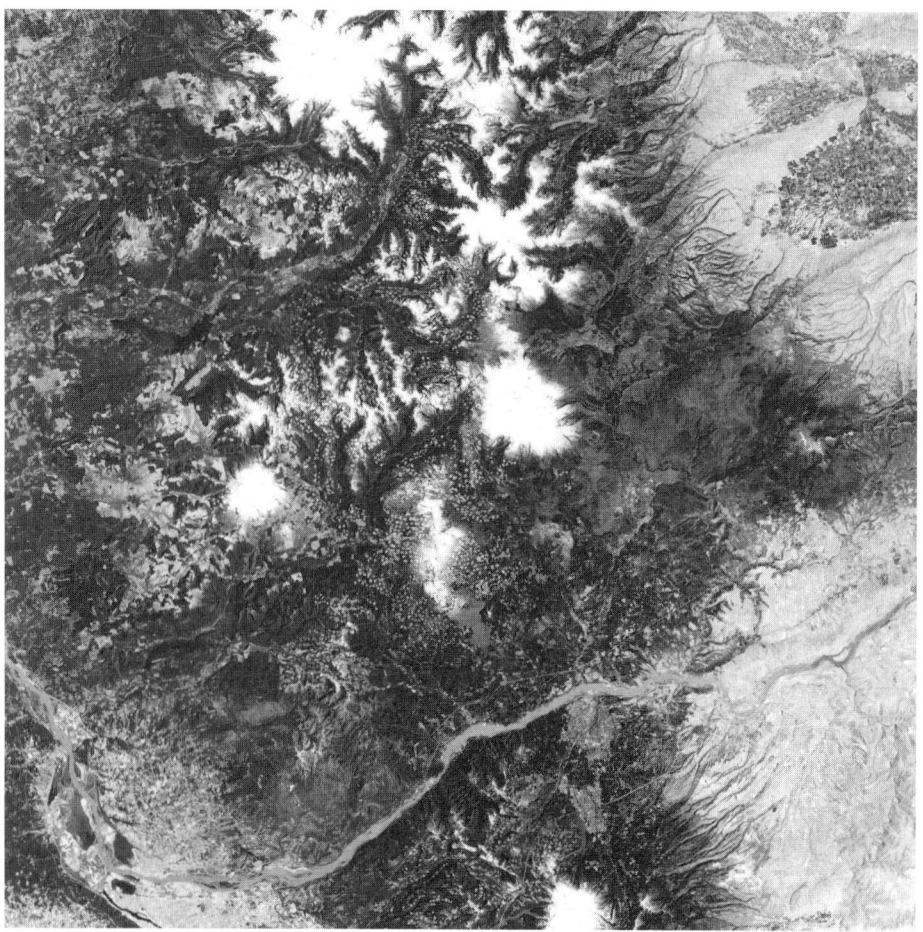

*Figure 5.17. Features change over time, as apparent in this series of Landsat images in the red portion of the spectrum. A very wet period is displayed by the size of the Columbia River and light grey tone of sediment in the river. Also, the wet conditions are evident in the dark tones of Sauvie Island wetlands in the lower left corner (June 30, 1974).*

Following the use of visual interpretation for change detection, digital methods began to be employed. Change detection methods have been broadly divided into either enhancement (precategorization) or postcategorization methods (Nelson, 1983; Pilon et al., 1988; Singh, 1989).

Image enhancement change detection techniques involve the transformation of two original images to a new single-band or multiband uncategorized image in which the areas of land cover change can be detected. The resulting image data can be further processed by other methods, such as by using a categorizing algorithm, to produce a categorized change detection product or thematic map of change.

These enhancement techniques or preclassification techniques were often based on

## Imagery and Interpretations

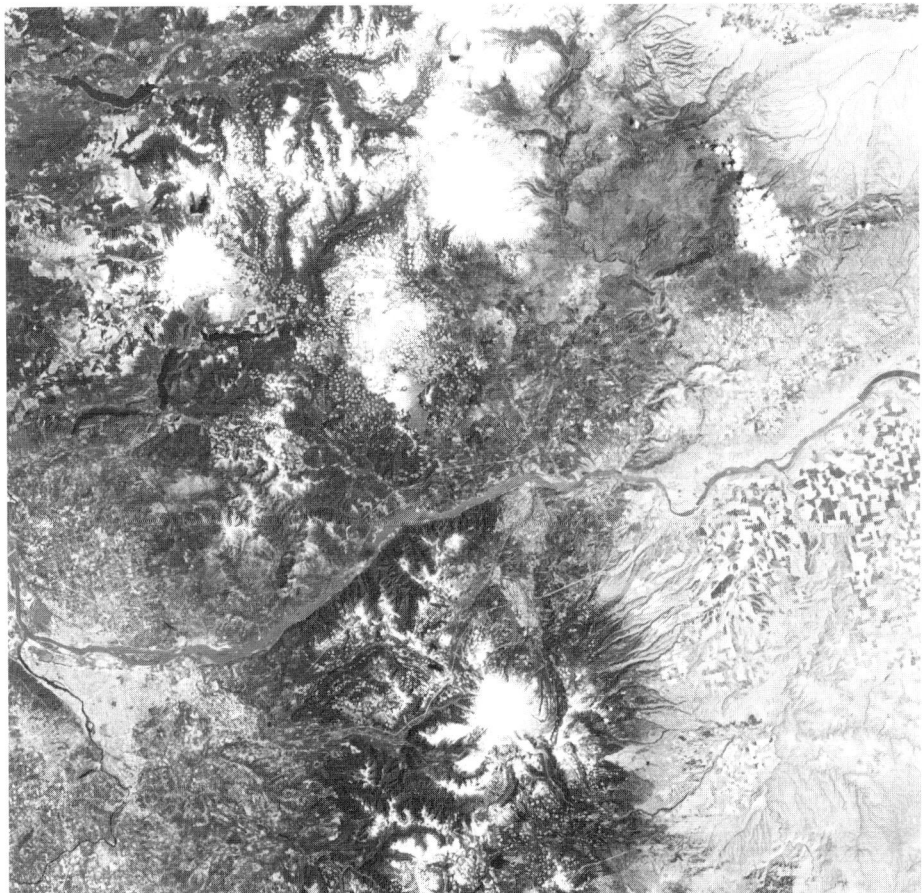

*Figure 5.18.* The same location as in Figure 5.17 is shown on May 29, 1975, or almost a year later. The larger extent of snow deposits are evident by comparison of peaks such as Mt. Rainier, Mt. St. Helens, Mt. Hood, and Mt. Adams north and east of Portland, Oregon.

the concepts of image differencing or image ratioing (Weismiller et al., 1977; Toll et al., 1980). Differencing of vegetation indices has proven to be a valuable approach for detection of change in sensor images (Lyon et al., 1998).

A valuable step is that of image equalization in the data preprocessing stage, which usually improves the results of change detection (Lunetta and Elvidge, 1998). Techniques like band-to-band regressions and principal components analysis have been used to simultaneously perform the image-to-image equalization and the detection of change areas.

In postcategorization change detection, two images from different dates are independently categorized. The area of change is then extracted through the direct comparison of the categorization results. An advantage of postcategorization change detection is that it bypasses the difficulties in change detection associated with the analysis of images acquired at different times of year or by different sensors.

The disadvantages of the postcategorization approach include greater computational and labeling requirements, high sensitivity to the individual categorization accuracies, plus the difficulties in performing adequate accuracy assessment on historic data sets (Lunetta et al., 1993).

Because all enhancement methods are based on pixel-wise operations or scene-wise plus pixel-wise operations, accuracy in image registration and coregistration is more critical for these methods than for other methods.

## Preprocessing of Satellite Data

One can use a set of procedures which will render high quality image scenes for derivation of ratio and index-type images, and for categorization of land cover types. The goal of applying these algorithms is to produce images with few system-related sources of noise. All or some of these procedures may apply to a given analysis or application, and the selective use of each is dictated by the type of sensor data and by the objectives of the effort. The procedures include smoothing of spectral variability, image compositing to remove clouds, image enhancements such as ratioing and indices, and others.

Processed images will often be smoothed or deconvoluted by "destriping" of images. This procedure removes the variability from row to row in the image due to differential response by one or more radiometric problems. In the past, Landsat Multispectral Scanner (MSS) detectors in the sensor array were found to drift from like calibration. A similar problem was noted in Landsat Thematic Mapper (TM) data sets. The variability appears as a stripe in early Landsat images, and is present in MSS data in general. Often there will be a general deconvolution of processed images, as it is often desirable to resample the pixel size to a known quantity (say, 57m × 80m for MSS data or 30m × 30m for TM data) and maintain geometric fidelity. This will also help to reduce the spectral oversampling in the cross-track or x-direction for MSS data, and potentially reduce other geometric errors associated with other sensor data.

Special efforts were made over the years to develop methods to create cloud-free composites of multiple date images of the same area (Loveland and Ohlen, 1993). This is necessary as some image scenes will be collected and they will exhibit clouds, and much work was conducted by the USGS EROS Data Center to develop these methods in support of the AVHRR biweekly composites program (Eidenshink and Haas, 1992).

The composite images are made of cloud-free portions of images from different dates, and processing procedures will also eliminate systematic and random variability. This variability results from changes in the sun's position, atmospheric influences, vegetation growth patterns or phenology, and other reasons. This variability causes the same materials or land cover types to exhibit dissimilar spectral signatures over time. A portion of this variability is reduced by calculating the normalized vegetation index (NDVI) using the red and near infrared portions of the electromagnetic spectrum. A processing system can be developed to check the NDVI values of identical position pixels in each image under study. NDVI values that are commonly associated with cloud-obscured pixels can be avoided, and only cloud-free pixel values can be written to the output image.

The composite images often will undergo additional processing to make the component images as similar spectrally as possible. Corrections of the scenes to a set solar elevation angle will establish similar solar illumination conditions. This can be effected by the image equalization procedures mentioned above and in Lunetta and Elvidge (1998).

The NDVI is used to make this valuable composite product. Many uses have been found for NDVI that illustrate the utility of biomass or greenness indices. For example, the calculation of the index or ratioing of image bands is valuable because it reduces the differential solar illumination effect. This occurs when mountainous areas receive more illumination on the sun-facing slopes than on the slopes facing away from the sun. Another valuable feature is that it also reduces atmospheric haze and other systematic noise contributions. These band ratios may be used as products in themselves or as input to categorization algorithms.

## The NALC Experience and Change Detection Procedures

The North American Landscape Characterization (NALC) project was conceived to enhance current and archival remote sensor data sets over a 20-year period to examine global change issues (Lunetta et al., 1993; Lunetta et al., 1998). The basic NALC data set consists of three dates of Landsat MSS data, from the early 1970s, mid-1980s, and early 1990s, plus ancillary data such as digital elevation model (DEM) data. In performing the change detection for the NALC project, two output products were generated: (1) land cover change from the early 1990s to the early 1970s; and (2) land cover change from the early 1990s to the mid-1980s. The output products should show the location of the changes as well as the nature of the change (e.g., conversion of woody to herbaceous and cover) over the approximately 20-year period (Lyon et al., 1998).

Image scene products have been developed and made available through the efforts of the USEPA and USGS EROS Data Center (EDC). The products resulted from development of "triplicate" scenes from the archival and current MSS database presented in whole-scene format (110 × 110 miles on the ground) with all the bands.

The NALC Landsat MSS triplicates were selected from the available archive of data held by the USGS EROS Data Center. A considerable effort was put into selecting scenes with low cloud cover and at or near the same date for all three time periods in order to have data sets optimized for digital change detection. However, in the extreme southern portions of the project area (southern Mexico, Central America, and Caribbean islands) it was frequently not possible to find low cloud cover scenes at the same time of year.

Dissemination of data was an important consideration of the NALC project. These products are available on media such as the Internet, CD-ROM, and can be browsed through communication networks. In addition, characteristics of these product scenes have been incorporated into USGS's database of satellite image scenes called the Global Land Information System (GLIS). This allows inventory and archiving of NALC products, as well as facilitating browsing, identification, and procurement of suitable products.

The NALC products can be obtained and processed by the user into enhanced products of interest. These enhanced images could include a greenness or biomass index image for each date of coverage, other ratio type images, and unsupervised categorization images with approximately 100 classes.

As part of the NALC effort, we evaluated a number of methods to enhance the data sets and methods to conduct change detection experiments. A lot of the knowledge and experience gained is provided here. Additional NALC and other experiential results can be found in the literature including the book by Lunetta and Elvidge (1998).

## Steps in Change Detection and Data Processing

The experiment begins with selecting multiple date data sets to identify the wetland or other features of interest, and assembling them in order to simplify the detection of change using digital image processing. Below is a description of some of the complicating factors which tend to make change detection more difficult, with indications of the design features of the NALC project which addressed the difficulties.

A number of characteristics of sensor data need to be addressed to achieve successful results. The spatial and spectral characteristics should be similar to facilitate analyses. If the data sets are from different sensors or bands in the electromagnetic spectrum, the resolution and spectral bandpass differences between images acquired with two sensors complicates the direct comparison or the digital analysis of the data to detect change.

With the spectral bandpass differences, the detectable land cover classes for the two dates of imagery may not be comparable. Land cover classes that are distinct when observed with one sensor may be indistinguishable with a sensor having broader bands or fewer numbers of spectral bands. If there are substantial spatial resolution differences between the two input images, ground features may be visible in one data set and undetectable in the other. In the NALC project we eliminated problems that could be introduced due to spatial resolution and bandpass differences by working exclusively with Landsat MSS, a sensor series that has provided relatively consistent spatial and spectral characteristics for a 20-year period.

Variations in the radiometric response of a sensor can complicate change detection by requiring some form of image-to-image equalization prior to the change detection or by requiring the use of a post-classification change detection approach (Lunetta and Elvidge, 1998). This factor is important to the NALC project because we are working with data spanning a 20-year period and collected by five different MSS sensors (Landsats 1-5). Although each of the MSS sensors was calibrated before launch, it is known that their response drifted over time and that there is no adequate way to calibrate the data to radiance units without extensive ground-based efforts. Similar problems develop when different sensors or different parts of the spectrum are evaluated that come from different sensors.

If clouds are present in images from one or both dates, it is impossible to detect land cover change between the two dates of imagery where clouds occur. To minimize the effects of clouds, low cloud cover was one of the primary factors directing the selection of scenes for change detection efforts.

Variations in solar irradiance, solar elevation angle, and solar azimuth will affect scene brightness levels and the location of shadows. The NALC project attempted to reduce these effects by selecting scenes for the three time periods which were acquired at or near the same date in order to match the solar conditions as far as possible. In general, high sun angles (low amount of shadowing) are better than low sun angles for the detection of land cover change.

Variations in atmospheric affects (scattering and absorption) can alter scene characteristics sufficiently that they must be considered in evaluating change detection methods. The NALC data sets have been visually screened to remove scenes where within-scene atmospheric variations are obvious, and now the use of "browse" files composed of metadata and compressed images can be used to select optimal images. By assuming that the atmospheric effects on the selected scenes are uniform across the entire scene area, it is possible to partially compensate for scene-to-scene variations in atmospheric effects by scene-to-scene brightness equalization (Lunetta and Elvidge, 1998). The coregistration of the image products will also help to compensate for topographic variations which alter atmospheric path length.

Phenological variations in vegetation result in large changes in the reflectance patterns of the land surface. If images of leaf-on and leaf-off conditions are compared, whole regions can appear to have been "deforested." The scene selection process of the NALC project has attempted to minimize this problem by selecting scenes from the same time of year for each of the selected time periods. In some cases it was not possible to do this due to the lack of appropriate archival data.

Spatial misregistration of images will tend to reduce the accuracy of any digital change detection effort. These effects are most severe on the change detection techniques using enhancement. The NALC project data sets were digitally coregistered, with accuracies on the order of half a pixel or less, creating data sets that can be analyzed for change with minimal errors due to misregistration (Lunetta et al., 1993).

## Change Detection Through Vegetation Index Differencing

The development of vegetation indices from spectral reflectance values is based on the differential absorption and reflectance of energy by vegetation in the red and near-infrared portion of the electromagnetic spectrum (Derring and Haas, 1980). In general, green vegetation absorbs energy in the red region and is highly reflective in the near-infrared region (Anderson and Hanson, 1992). A number of vegetation indices have been formulated and utilized for monitoring vegetation change. Of these vegetation indices, NDVI has been used most widely for monitoring terrestrial vegetation dynamics (Townshend and Justice, 1986; Eidenshink and Haas, 1992; Tappan et al., 1992; Lyon et al.,1998). The NDVI compensates for some radiometric differences between images; however, it does not completely remove radiometric noise from images that are being compared. The difference in the NDVI values of two images in certain cases responds to changes in land cover (Nelson, 1983; Banner and Lynham, 1981). Singh (1989) concluded that NDVI differencing was among the most accurate of change detection techniques.

In change detection pilot data studies, the NDVI values were computed for each date of imagery (Lunetta et al., 1993; Lunetta and Elvidge, 1998; Yuan et al., 1998). The early 1970's and the mid-1980's NDVI images were subtracted from the 1990's NDVI image to create two NDVI difference images. These results were compared to the photointerpretation results in order to calculate the accuracy of the NDVI differencing approach (Lyon et al., 1998; Yuan et al., 1998).

## Change Detection Through Principal Components Analysis

Principal Components Analysis (PCA) involves the reorientation of axes of an input data set, creating output Principal Components (PC) data sets. In the case of change detection, two coregistered images would be input as, say, an eight band or eight axis image. Because of the autocorrelation of the original data, there is an elongated "cloud" of distribution of data located in the axis of each data set (Lunetta and Elvidge, 1998). The PCA will orient the first axis of the output data through the central core of the input data cloud such that the interband variability is maximized. The second axis will be perpendicular to the first and will be directed through the next major direction of variance in the data set such that the intraband variability is maximized for the second axis. The creation of new axes continues until a default (eight in this case) or limit is reached.

The first Principal Component explains the major variability in the image and contains the overall scene brightness variations which are in common between all the input bands. The second, third, and fourth (and sometimes higher) PC images frequently contain information on pixels which changed in reflectance between the two dates of imagery (Byrne et al., 1980; Richardson and Milne, 1983; Fung and LeDrew, 1987). The last PC image would be expected to contain random noise that existed in one image relative to the other.

In scenes with cloud cover and associated shadowing, it will be necessary to screen out the cloud and shadow areas to exclude them from the PCA (Lunetta and Elvidge, 1998). Clouds or shadows which are present in one date and absent from the second date will tend to redirect the axes which would otherwise be established due to reflectance changes due to land cover change.

Part of the PCA algorithm involves the normalization or equalization of the input data, thereby reducing atmospheric and sensor radiometric response differences from the output PC images. This is one of the characteristics required for many change detections, making PCA a useful approach in certain analyses.

For interpretation of results, one can make brightness/greenness images using the first three Principal Components. Developed over several years, the brightness/greenness images facilitate interpretations of cause and effect, and can help identify the location and the spectral characteristics of changed land cover types.

## Postcategorization Change Detection

Postcategorization change detection involves the categorization or computer classification of images, and labeling of land cover thematic classes from each year using the same class types from the same classification system. The locations of change can be identified by the areas of change in land cover from the earlier date image to the later date image.

As in the preclassification change detection methods, the input image data can be optimized to remove sources of error that are spectral and geometric. The examples discussed above include sensor noise (e.g., stripping), atmospheric effects or haze, and differential solar illumination. After data enhancements, the input images can be used in categorization.

The data are to be processed by steps that include image preprocessing, training set selection, training set evaluation, and maximum likelihood categorization. Image analysts or cooperators can identify the clusters or classes of the Preliminary Land Cover (PLC) maps, aggregate and assign land cover types from the categorization system, and create the Land Cover (LC) products.

## Training Set Development

The development of training sets for categorization requires a method that will yield high quality and unbiased training set characteristics in an automated fashion. The unsupervised training set development approach has been proven successful in providing good, homogeneous spectral clusters in applications such as land cover or landscape characterization. The unsupervised approach negates many systematic problems associated with atmospheric and terrain radiometric distortions. This approach has been used successfully for over 20 years, and it can also be implemented with a minimum of effort. Consistent results can be expected for most applications.

Unsupervised training set development is conducted with an automated spectral clustering algorithm. A common clustering method can be selected from the available choices in the literature of remote sensing applications. Suitable examples have often been implemented on applications software and hence are available to most users.

Training set development can be conducted with some assumptions and input parameters. In change detection projects, individual images under study should be subjected to unsupervised training set selection or clustering separately. Each image scene can be clustered on an iterative basis for as many as 100 or more clusters or classes. Criteria for linking or breaking of clusters are usually established empirically, and the defaults can be used initially until experience is developed as to the most appropriate conditions for the given application.

The results of unsupervised training set development can be evaluated in several manners. A preliminary categorization image will be produced from the unsupervised training set development or clustering activity. This image will help indicate the location of many clusters or classes. To both identify the type of land cover associated with each class, and to evaluate quality, the preliminary categorized image can be viewed individually class by class. Each class can be colored from a table of colors to facilitate analysis.

The preliminary categorized image can be used along with other procedures to evaluate the quality of individual training sets or clusters using aerial photo or other data. Procedures include divergence calculations, bivariate scatter plots of cluster means and variance-covariance, and comparison of the preliminary categorized image with ground information or aerial photographs.

Following analyses of the quality of training sets, the contents may be edited to remove examples of poor clusters, e.g., clusters that represent very few pixels in the original image, overlapping clusters, and clusters that are composed of more than one distinct, spectrally homogeneous land cover class.

## Categorization

The edited training sets should be used for categorization of original scenes. The maximum likelihood or Bayesian decision rule approach to categorization has been evaluated. There are a number of reasons to use this approach, among which is the fact that it has been used successfully in remote sensing applications for over 20 years. This approach makes use of both Euclidean distance measurement and probability-based criteria in the determination of the class identity of a given pixel. The use of these criteria makes the categorizing algorithm more sensitive to class characteristics as compared to single-criterion approaches (Jensen, 1996).

After evaluation of training set quality, the image scenes can be categorized using four bands as input to the maximum likelihood classification algorithm. A sensitivity experiment or feature selection (Moik, 1980) can evaluate the appropriate bands to be employed. Studies have shown that one should employ only those bands important to identify the land cover types of interest. One can select a subscene of the data of small size (say, $250 \times 250$ pixels). From the literature, a set of several bands can be selected and this small subset categorized. Comparison with aerial photos allows determination as to the value of the bands that were selected. Iterations of different combinations of bands through the categorization algorithms allows the selection of the optimal combination to identify the land cover types of interest (Lyon et al., 1992).

The resulting product of 100 or more classes will represent a Preliminary Land Cover (PLC) product. It remains to identify the land cover type of each cluster or class. This activity is known as "labeling" and can be performed by using local knowledge, ground information, aerial photographs and maps. The PLC product often requires that certain clusters or classes be aggregated or "lumped" to render the best product. This aggregation will allow inventory of several spectrally distinct clusters or classes that may represent the same functional land cover class as represented in the selected land cover categorization system.

## Land Cover and Classification System

The land cover classes in the PLC image can then be labeled as to cover type using a land cover classification system. Classification systems were developed to specifically support project objectives, and to be compatible. Most classifications can be interpreted with the other major land cover classification systems.

Any custom land cover classification system should be compatible with the Anderson et al. (1976), Cowardin et al. (1979), Brown et al. (1979), and/or other commonly used (like Omernick) and accepted systems. These systems have been optimized for inventory applications that include wetland land cover types. This greatly facilitates understanding by later users.

The assignment of class types to PLC classes should be performed by personnel familiar with these techniques and experienced in this practice. The work may also have been done with the help of local cooperators familiar with the general land cover and wetland land covers of the study area.

The level of detail of classification system can be selected by client groups. Level 2 or level 3 is often used as the final level of land cover type detail to be identified. This approach will supply sufficient detail while using the capabilities of remote sensor data to their best advantage. It is recognized that differentiation between wetland forest types and wetland shrub/scrub classes at Level 2 will be difficult and in some cases not possible without the aid of ancillary data. However, a best effort will be made to accomplish this sort of differentiation within the requirements of the given application project.

The final categorization images should be accompanied by details on the training sets including cluster or class means in each band, variance-covariance matrix, class identities, and details related to quality and accuracy.

The resulting product will be a land cover thematic data coverage product (LC) for each scene. It can then be supplied to the client groups for validation. Subsequent to data validation, LC products should be archived for reference and later use.

## Assessment of Accuracy

The assessment of accuracy of the categorized or classified product is a necessary step to ensure a quality product for subsequent analysis, and to meet quality assurance and quality

*Figure 5.19. Low altitude aerial photographs are acquired to make large-scale topographic maps for engineering of highways. The state department of transportation or like agency archives these photos, which supplied fine detail on wetlands such as these in Trumbull Country, Ohio.*

control (QA/QC) objectives. This can be accomplished by fieldwork and/or photointerpretation of the land cover classifications, and conducted according to the scope or guidelines by a team of photointerpreters and/or field personnel. The photointerpretation/field team should utilize an accepted land cover classification system as specified in a scope or a call for proposals document to verify the results of categorization.

Photointerpreters should utilize a number of other sources of data. Some of these data sources are included in most studies. The data sources can include color infrared high altitude photographs, Landsat and/or SPOT Image data, aerial and map atlases as available, and USGS quadrangle maps and digital files.

In addition to these sources, the team should utilize any other sources available to the researchers. These include: USDA National Resource Conservation Service (NRCS) soil surveys; USFWS National Wetland Inventory (NWI) maps; available USGS orthophoto quadrangle products; maps of Federal property boundaries; and other sources.

Local groups or companies often have a number of complete and partial county coverages for the area to be studied. These archival photographs can be used along with other data sources to map and classify land cover according to a given scope. These coverages are often at medium and low altitude, and they supply a lot of detail that can support interpretations using high altitude photographs or satellite products as the base data source. These photographs represent a very valuable resource that local groups bring to the project, and use of these detailed and contemporary photos will both speed the classification work and enhance the accuracy of the type mapping.

Satellite data such as Landsat or SPOT Image data can also be processed using image processing systems to produce additional images to assist in interpretation and accuracy assessment efforts. Image enhanced products can be used to assist photointerpretations, and to help visualize local and regional land cover.

Computer categorizations can be made in a few local areas where data of that given resolution can be of value. It is anticipated that ancillary satellite image data can be helpful in classification of land cover for such themes as wetlands and agricultural types (Figure 5.19). Satellite data will be most useful at level 2, and certainly will be valuable at level 3 of a given land cover classification system.

These quality determinations should also be made after the final digital databases or land cover products are produced. In this manner the final polygons and classifications can be checked against the original photographs, supporting data sources, and the original interpreted overlay.

## Applications to Wildlife Habitat Quality Evaluations

A number of researchers have worked on habitat quality ratings for purposes of modeling wildlife resource conditions. These results have implications for addressing impact on resources in the EIS or EA process. Often taking a census of the actual animal populations is impossible. They may be too rare, the populations may fluctuate naturally over multiple-year periods, or conditions may be too hostile for human sampling of animal populations.

Use of habitat rating models such as USACE HEP and other models allows the evaluation of current or preconstruction conditions, and future project or activity conditions. Habitat models allow "gaming" or the evaluation of scenarios as to the effect of alternative land and resource management plans. This allows evaluation of impacts of alternatives or different management plans.

# 6 GIS Applications

## Background on Geographic Information Systems (GIS)

For a number of years people have developed software and hardware tools to address the spatial characteristics of data. These tools are known as Geographic Information Systems (GIS) in a general sense, because they address data that has a spatial component, and the spatial and other characteristics are manifest in the storage of data.

GIS systems have the capabilities to store and process data, operate on data sets using algorithms or models, and to present the results of these transformations in the form of maps. GIS systems can be employed for a number of activities of specific nature in terms of transformations including repeated evaluations of scenarios or simulations, data and results storage, and visualization of results.

One can think of GIS as a series of files that contain information of varying types. The information is stored "layer by layer" in the same spatial format. This allows one to compare and contrast different information for the same spatial locations (Figure 6.1). Integration of information or data over an area of interest allows for area-based data analysis.

Data and/or information can be stored in at least two different manners. Some data types are best represented as a series of grid cells, also known as matrix or raster data format. This is the same method that is used to store arrays of data in computers. Grid cell format can also be processed mathematically by linear algebra, and all transformations can be represented as matrices and operations on matrices. Each grid cell holds a value that represents an actual measurement, a real or integer number, or a categorical variable such as a code number or letter of a type class or attribute.

A second storage method is vector or coordinate system storage, where the (x,y) coordinates of a feature are stored along with code that describes points, line positions, polygon shapes, and other characteristics of spatial features called topology. Identification of characteristics of the point, line, or polygon are made by assigning a name tag or attribute. Attributes may indicate the type of feature described by the point, line, or polygon, the characteristics of a variety of features associated with the polygon, and the adjacent characteristics of the polygon or line segment.

The GIS is valuable as indicated by the fact that one can operate on the detail stored in

*Figure 6.1. Integrating the various processes that occasion a wetland requires the capability to store and process multiple variables, and to model from the resulting GIS database. Complex areas such as the Great Salt Lake Wetlands need detailed data and integration.*

the database, and use these data to address an application of interest. The variable or data "layers" in the GIS support analyses of applications using the appropriate characteristics germane to the wetlands or other ecosystem under study.

## The Utility of GIS

Wetland location and enumeration efforts are also known as wetland inventories, and inventory information is a basic requirement for managing ecosystems and for evaluations of risk. Often this is the first and most important goal for generating a GIS for a given area. As the first goal is met, further goals and objectives are commonly defined to make use of the inventory data and the utility of GIS.

An important capability of GIS is the simulation of physical, chemical, and biological processes using models (Figure 6.2). The advent of landscape ecology and landscape characterization studies has greatly advanced the use of ecological indicators that can lend insight into processes. GIS and remote sensor data can greatly facilitate landscape characterization studies of wetlands.

The current need for information often requires the use of models. Improvement in GIS and related technologies has dictated the advancement of GIS technologies to integrate suitable algorithms or models that apply GIS to statistical and deterministic modeling efforts. Many previous efforts have used commonly accepted or traditional modeling or analysis methods that preceded the advent of GIS, and have achieved partial success. Much effort is ongoing to join the approaches and optimize their capabilities.

GIS can potentially be used with deterministic or complex models based on algorithms that simulate processes, or they can be applied to statistical models. The requirement is that the model to be applied has the capability to take spatial and/or multiple file or "layer" data as input to computations.

The characterization and management of wetland ecosystems and receptors by personnel can be improved with the integration of computer, sensor, and GIS operations. From a database management "style" of system, office and field people would have certain basic information available for their work.

*Figure 6.2. Great Salt Lake wetlands have the added complexities of saltwater systems and of freshwater runoff from snow melt, with a background of rural to dense urbanization. These processes and stressors almost necessitate the use of GIS for analyses.*

For work in the field, the data on wetland characteristics can be studied by office and field personnel. Each can obtain hard copies to use, or view results on a field portable computer. Field checks can be made and placed on hard copy or in computer files while examining the location of interest. Map sheets or images generated from high resolution satellite sensor images or digitized aerial photographs or orthophotographs can also be made available for fieldwork.

Results from GIS and remote sensor data can also be evaluated as to quality. Quality Assurance and Quality Control (QA/QC) procedures can be used to determine the quality characteristics of the product, and to assess the accuracy of the product. The format and spatial characteristics of GIS and remote sensor data assist in the evaluations of accuracy of the input data themselves, determinations as to sources of uncertainty, and for assessments of methods used in experiments and the results.

## Data Sources and Their Application

A hardship in creating a GIS is the acquisition of data to populate the thematic layers. It is very desirable to take advantage of existing data sets, as well as to develop data particular to a given application and store them in the GIS for analyses. Data input may represent as much or more than 60% of the initial capitalization of a GIS system. Industry has recognized this need and commercial opportunity, and now makes data sets available for many areas at reasonable cost.

To facilitate analyses by computer and other methods, all appropriate data should be transferred to a database or GIS type system. Identification and monitoring of features can be accomplished using remote sensor, map, resource database, and other inputs to GIS.

The available data sets take many forms. Many record the presence and location of features across the landscape. These data are stored digitally, and they can be incorporated into a given system. Existing data sources may include historical maps, historical aerial or ground photos, internal or "grey literature" reports with spatial data, and any other appropriate data suitable as data inputs.

Historical maps can be of value in developing data layers and evaluating the historical characteristics of the wetland ecosystems or wetland hydrology. The U.S. Geological Survey (USGS) and National Archives and Records Service (NARS) maintain original copies of maps. They provide services to supply copies of these archival maps. The copies can be rendered in digital form by scanning or digitizing, as appropriate.

Current map products are available for land areas from the USGS. River, large lakes, connecting channels, harbors and ports, coastal and marine areas are documented on maps supplied by the National Oceanic and Atmospheric Administrations (NOAA).

Often a wetland is found in an older urban area that has been a factory district or industrial area in the past. To characterize past activities in urban areas, Sanborn Maps (Pelham, NY) can be contacted to determine the existence of their fire map products. Sanborn maps were used historically to record the characteristics of industrial buildings and their contents for insurance purposes. The maps may be useful in identifying historical fill areas, potential deposits of hazardous waste products, and other details of human activities (Lyon, 1987).

There are many vendors that can supply digital map data to assist in the development of data layers. Often the vendors start with a government database and add value to it by supplying updated information and additional information.

One useful tool is the U.S. Bureau of Census Topologically Integrated Geographic (TIGER) data or vendor enhanced versions of these data. TIGER data products hold census information and related details, and Zip Code information. These data can be valuable in evaluating population conditions adjacent to wetlands to assess potential exposure and risk to wetlands. They can also be useful in studies of the demographics of a given wetland area, for evaluation of recreational management, and for planning purposes.

Another useful source of data is those products created by the U.S. Geological Survey (USGS). These products may also be enhanced by vendors and supplied in a customized form to help the development of GIS.

GIS Applications

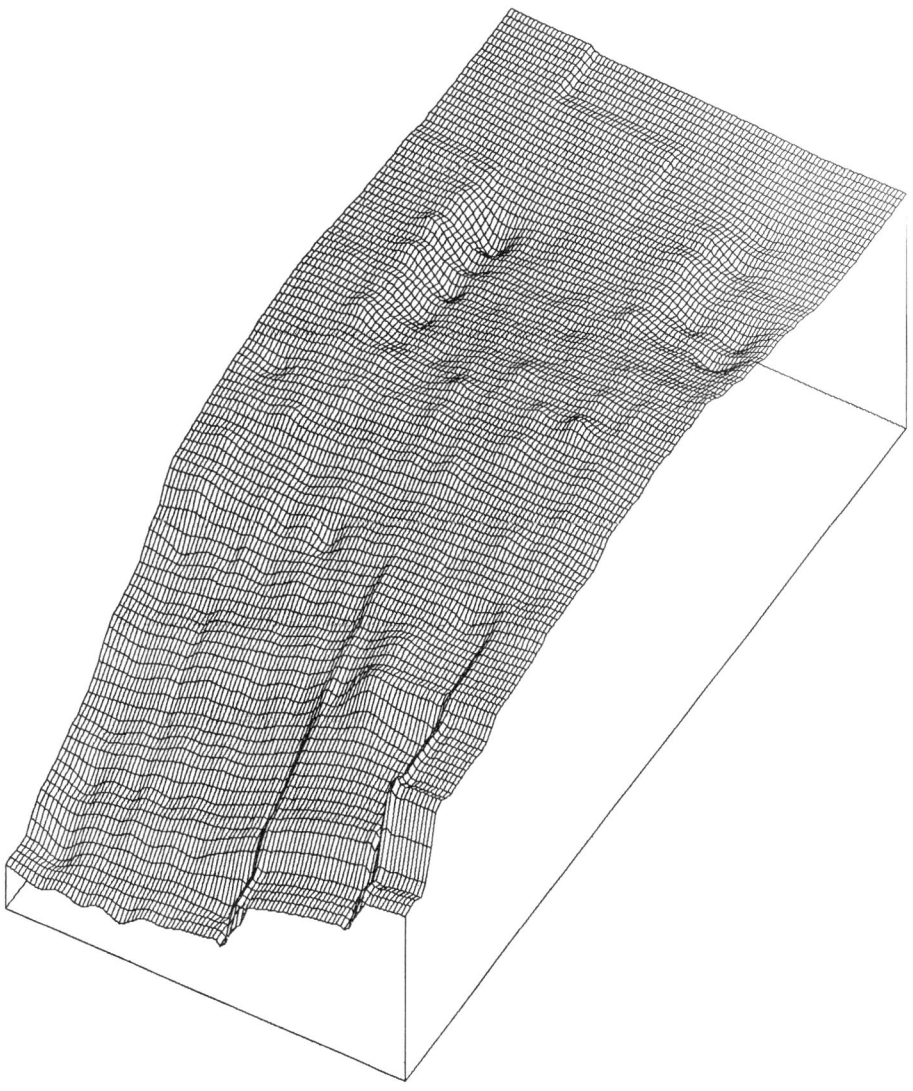

*Figure 6.3. Digital databases and GIS can do simple and complex jobs. Pictured here are wire-mesh graphics of wetland variables including relative elevation of a coastal wetland area in the Straits of Mackinac, Michigan.*

An example data type is the Digital Line Graph (DLG). The DLG holds some of the information that composes U.S. Geological Survey maps. These data are stored in vector format, by theme or attribute type. The DLG have thematic data or attributes found on most USGS quadrangle maps at a scale of 1:24,000 or 1:100,000. Themes include natural and human-related features such as topography, stream drainage, road networks, homes or buildings in rural areas, and other details.

The USGS and other vendors also create point elevation data in grid cell format, known as Digital Elevation Models (Lyon, 2000). DEM data are very valuable inputs to GIS by them-

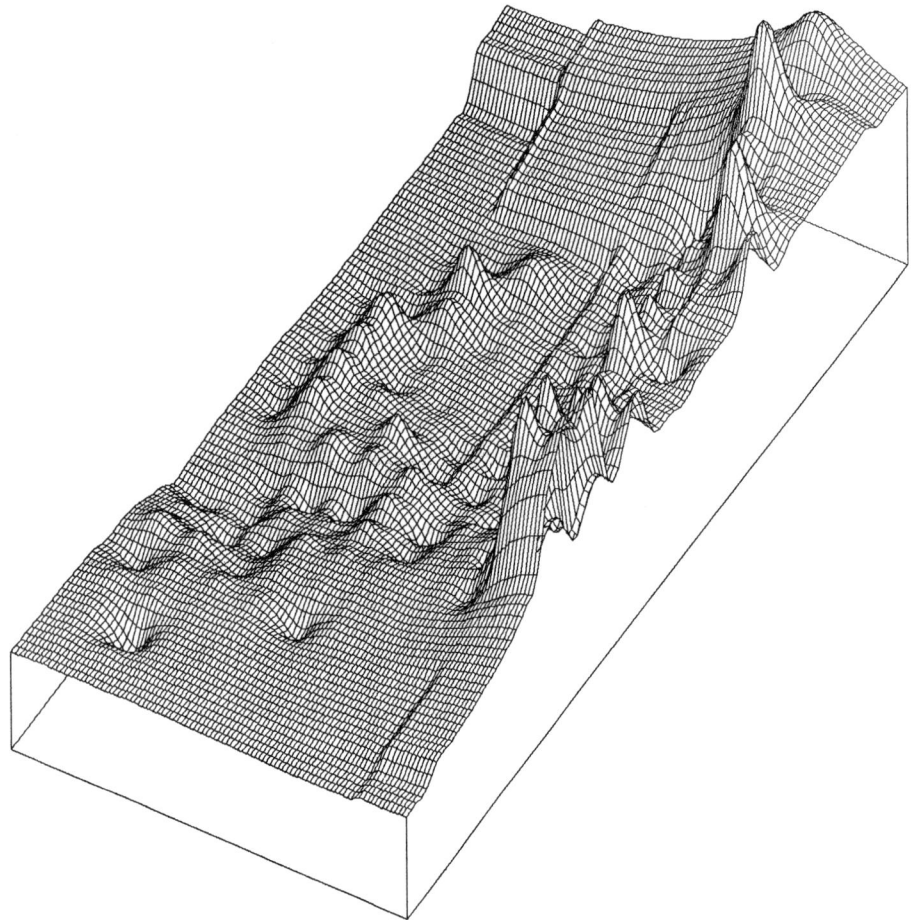

*Figure 6.4.* Processes can be displayed in the same format for comparison such as duration of flooding during the vegetation growing season, as caused by Great Lake water level fluctuations.

selves, and can be used to develop contour maps of the study area (Figure 6.3). Additional data processing can yield valuable hydrological information. One can develop information as to watersheds and subwatersheds, routing of water (Figure 6.4) and waterborne materials (Figure 6.5), and hydrologically important topological data.

A valuable data set for study of wetland and other ecosystems is the digital soils data available from the U.S. Department of Agriculture (USDA). The USDA Natural Resource Conservation Service (NRCS) has created general soils data in several forms, and products are available at different scales and levels of detail concerning soils. One example is the STATSGO data sets which provide general soil information suitable for watershed and large area analyses.

These USGS, NRCS, and other governmental products have the potential for being inaccurate. This may be due to systematic or nonsystematic errors, land cover changes since

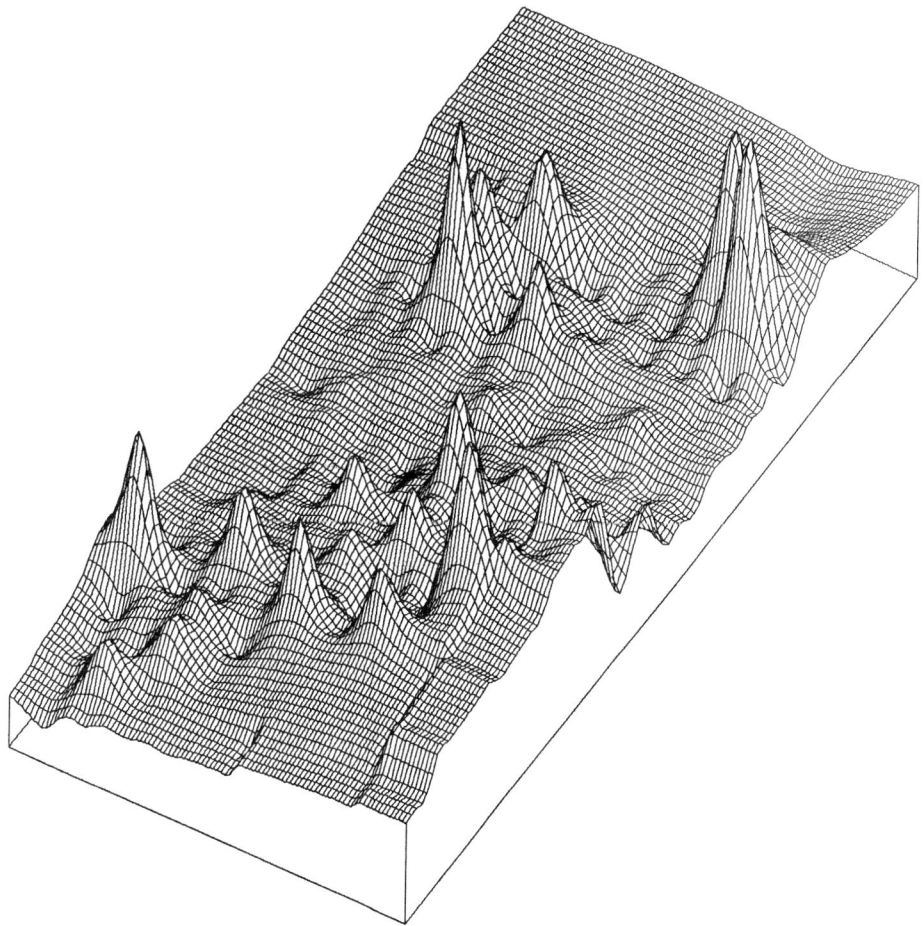

*Figure 6.5. The change in variables resulting from processes can be visualized, and used to evaluate stressors and their exposure to receptors such as wetlands. Pictured here is the spatial distribution of concentrations of macronutrients, in this case phosphorus.*

their production, or other considerations. Often the data require some preprocessing to be implemented. Preprocessing may include corrections or additions of data or attributes, rectification of map coordinates, and other transformations. The user will make any corrections or modifications for the needs of a given project, or purchase value-added products from a vendor who has accomplished the corrections.

## Wetland GIS Applications

A number of wetland GIS applications have been completed over the years, and they nicely illustrate the potential of these technologies. Many of these efforts have resulted from the need to address difficult to achieve project goals. GIS applications are applied due to the variability of wetland and related ecosystems over time and space, and the number of variables that must be evaluated.

GIS will allow the integration of remote sensor data and the available database information. Use of a combination of technologies allows the detection of the presence or absence of wetlands and other receptors, or the influences of human activities in the area of interest. GIS provides the capability for integration of the landscape and other characteristics of the wetland with imagery on a given, localized study site. This combination will contribute to management functions including inventory, monitoring, planning, and risk assessment.

The capabilities of using GIS for wetland studies have been identified through a number of efforts. It is evident from the wealth of contributions that GIS technologies are a potentially good solution for many applications. In general, GIS brings a lot of utility for evaluation of wetland ecosystems. GIS allows the integration of the variety of variables that occasion the presence of wetlands (Lyon and Adkins, 1995). GIS allows users to integrate hydrology, soils, and water depth data to identify why certain wetland plants or plant communities are found where they are, and to do so on a scientific basis (Lyon et al., 1986; Williams and Lyon, 1991; Jensen et al., 1993). From this sort of knowledge it is then possible to predict the presence or absence of general wetlands based on the needs of various management scenarios.

## Aerial Photo Analyses of Historical Wetlands

The Laurentian Great Lakes occupy two-thirds of their own watershed, and are very responsive to precipitation conditions from year to year. The Great Lakes can experience a six-foot range in water level elevations over time, making the fluctuations in water level a forcing function or stressor. The Lakes have gone from very low levels to very high levels and vice versa in as little as three to four years. This variability has caused Great Lakes wetland and limnology researchers to emphasize multiple date measurements of wetlands to best capture the variability in quantities of wetland receptors, and to develop the best data sets to facilitate evaluations of causative stressors or forcing functions.

The implementation of GIS has allowed for detailed evaluations of wetlands over time and space. An early effort in the connecting channels of Lake Superior and Lakes Michigan and Huron demonstrated the power of GIS (Williams and Lyon, 1991). Five, and later eight, dates of aerial photographs were interpreted by USFWS contractors (Wilen, 1990) for types of wetlands, and the quantities of each were described using the USFWS wetland classification system (Cowardin et al., 1979). These interpreted boundaries were converted from vector to raster files, and later processed on a commercial image processing/GIS system.

Year to year digital comparisons of general wetlands in the St. Marys River connecting channels allowed a number of evaluations to be made, including the abundance and type of wetland receptors during high water years as compared to low water years, comparisons of years with winter-season navigation stressors in the presence of ice versus non-navigation years, and other combinations of natural and human-induced stressor and exposure conditions (Lyon et al., 1994).

The GIS allowed for a large area to be evaluated which covered parts of seven 1:24,000 scale USGS maps. It kept track of every grid cell element of wetlands, and allowed comparison of presence or absence of wetlands over time, and any change in type of wetlands. The resulting detail supplied information on both the natural variability of coastal wetlands and fluctuating water levels, and any impacts of human activities or exposure (Williams and Lyon, 1991, 1997). The resulting graphic products were used in an Environmental Impact Statement (EIS).

The above study made one of the first uses of multiple date National Wetland Inventory (NWI) digital files for analyses. Similar digital NWI products are now available for use in GIS studies and can be valuable as an additional layer of data describing general wetlands. Digital files from the NWI are usually available from one date of aerial photographic coverage only, yet they can provide a useful data set for identification of potential jurisdictional or general wetlands (Lyon, 1993; Lyon and McCarthy, 1995).

## Use of Remote Sensor Data and GIS

Many applications make use of remote sensor data as input to GIS applications. Use of remote sensor and other available data sets may also present a true opportunity to leverage the capability of GIS. This is because a fundamental cost of starting and operating a GIS system is data input. If the available data sets are suitable to company or group applications, incorporation of these free or cheap data sets can represent a true savings.

Several studies have evaluated the capabilities of these sensors for general wetland analyses (FGDC, 1992; Lampman, 1993; Lee and Lunetta, 1995; Lunetta and Elvidge, 1998). Satellite and aerial remote sensing data can supply good information on the extent and variety of general wetlands, and detailed information such as general wetland quantity and type (Figure 6.6), biomass (Gross et al., 1987), and presence of lake macrophytes (Christel-Rose and Scarpace, 1991; Jensen et al., 1992; Welch et al., 1992; Remillard and Welch, 1992, 1993; Jensen et al., 1993).

## Applications to Marine Coastal Wetlands

Coastal areas of oceans, seas, and large lakes have received increased study for a number of reasons. These areas are at the edge or interphase between terrestrial and open water environments. As such they present a mix of characteristics, and serve a number of functions and uses. A large number of people inhabit many coastal areas, as do a number of lifeforms during some or all of their life. This results in complex systems, and complex demands on

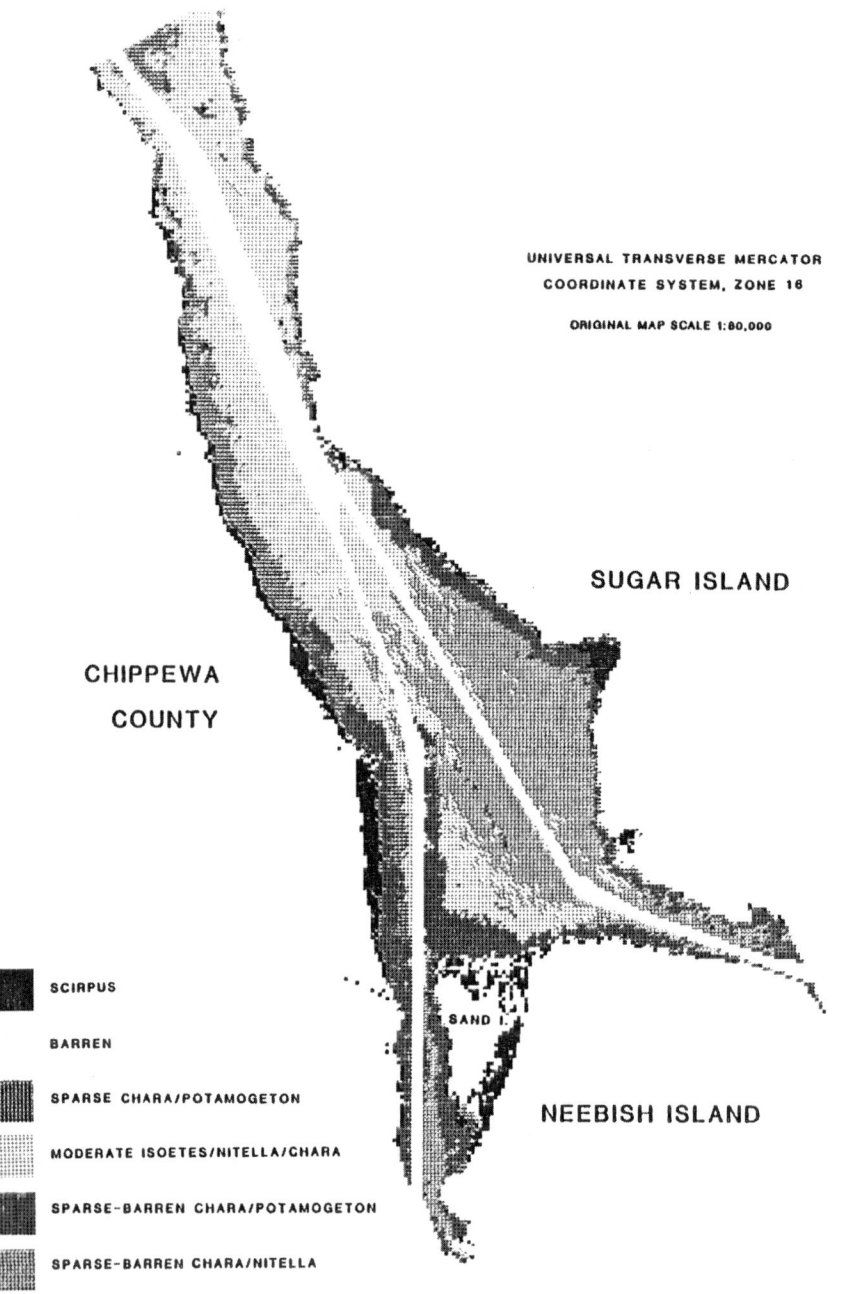

*Figure 6.6. Aquatic, submergent wetland vegetation as developed from a GIS system that modeled the distribution of plants based on bottom type and water depth.*

those systems from humans and biota. Remote sensing and GIS have proven to be valuable technologies to generate data, organize data, and evaluate management scenarios.

Because coastal areas of large lakes and marine areas contain wetlands, many of the applications previously addressed are pertinent here. There are a number of high quality applications that have been completed in coastal zone areas. These involve a variety of applications, including assessment of wetland landscape characteristics, land cover, land use, human stressors and risks, wildlife habitat, and global-change-induced flooding (Lee et al., 1992).

Satellite and aerial remote sensing data can supply good information on the extent and variety of coastal wetlands, and detailed information such as presence of kelp (Augenstein et al., 1991), and presence of mangrove and swamp vegetation (Lo and Watson, 1994; Ramsey and Jensen, 1995, 1996).

The combination of GIS and remote sensor technologies has proven useful in coastal studies (Miller and DeCampo, 1994). The addition of these technologies along with traditional on-site measures (Paul and Morrison, 1995) provides an excellent approach to measure and model highly distributed ecosystems. The presence of tides makes for more dynamic hydrologic conditions, and GIS is a great help to address the complexity of a number of on-site measurement requirements that are often encountered (Pearlstine et al., 1993).

Tropical and semitropical coasts often have colonies of mangroves that supply many wetland and habitat functions to the coastal zone. These areas are difficult to visit and measure in a complete sense without the use of remote sensor data in combination with on-site sampling (Ramsey and Jensen, 1995, 1996).

Evaluations of other types of coastal vegetation have been successful using a combination of remote sensing and GIS technologies. In particular, kelp beds on the Pacific coast and submerged aquatic vegetation (SAV) on the Atlantic and Gulf coasts have been measured in great detail and analyzed using remote sensing, GIS, and on-site sampling.

The implementation of radar remote sensing (Figure 5.14) can supply additional details on coastal and other wetland resources (Lyon and McCarthy, 1981; Ramsey, 1998). A useful application is the inventory and characterization of mangrove areas including tidal flooding conditions. Use of remote sensing and GIS database capabilities allows the detection of change due to meteorological events such as hurricanes (Ramsey and Jensen, 1996). In particular, radar remote sensing, traditional remote sensing, and GIS hold great potential for addressing complex coastal issues (Lyon and McCarthy, 1981; Hussin and Hoffer, 1989; Wu, 1989; Dobson and Bright, 1993; Klemas et al., 1995; Ramsey and Jensen, 1996).

The advent of Coastal Zone Management statutes and regulations has provided opportunities for increased study of coastal areas and their wetland ecosystems. It has also provided additional regulatory mandates to be addressed. As such, GIS and remote sensing have helped to organize coastal zone data, and help supply numbers for management and planning (Welch et al., 1988; Welch et al., 1992; Dobson and Bright, 1993; Klemas et al., 1995).

A number of efforts have involved monitoring of change in coastal areas (Niedzwiedz and Batie, 1984) or the potential for change as a form of environmental risk (Lee et al., 1992). A concern is the potential influence of global change on sea level and possible flooding, and the general influence of wetlands on flooding (Novitzki, 1979). GIS allows the evaluation of these future scenarios such that contingency plans can be developed for the coastal zone (Breininger et al., 1991).

Estuary areas are of particular interest in coastal zone management. The mixing of the runoff from terrestrial ecosystems with marine systems provides a very productive environment. It also produces many conflicts between natural and human activities. GIS and remote sensing technologies offer great potential for study and management of these areas. Past work has demonstrated the capability to track the changing characteristics of estuaries using on-site and remote measurements (Bagheri and Stein, 1992; Brondizio et al., 1996). GIS provides the potential to store and analyze all the variables necessary to understand these exposures and evaluate risks (Donoghue et al., 1994; Friel et al., 1995).

Coastal zone wetlands are influenced by the freshwater runoff from terrestrial ecosystems. Often, the runoff includes sediment and other nonpoint sources of pollution. These phenomena have held problems for a number of coastal areas and wetlands including Chesapeake Bay (Ackleson and Klemas, 1987).

Nonpoint source studies are greatly facilitated by the use of remote sensing data on land cover. Land cover data incorporated into runoff, transport, and other models in GIS helps to study these systems (Lyon et al., 1988; Jakubauskas et al., 1992; Kang and Bartholic, 1994). The goal is to identify potential remedial actions and halt the influence of nonpoint pollutants on coastal waters.

## GIS Analyses of Large Lakes and Lacustrine Wetlands

Lake or lacustrine ecosystems are often difficult to evaluate as compared to terrestrial ecosystems. Access to the resource generally requires the use of boats or ships, and the sampling of variables is most often a three-dimensional problem due to depth of lakes. Boat and ship time can be expensive along with personnel, and the processing and storage of data argue for advanced methods to facilitate analyses. Lake conditions can change rapidly with time as compared to some terrestrial ecosystems, and are very responsive to wind and other weather conditions. This often results in requirements to obtain more samples than may be necessary for study of terrestrial ecosystems.

The advent of GIS and of advanced remote sensor systems has greatly assisted the evaluation of lacustrine or lake-related ecosystems (Samuels, 1993; Lyon and Adkins, 1995; Troge, 1995; USEPA, 1995). Boats and ships can only be present at one location at any given time. The synoptic view of satellite or aircraft remote sensors allows for simultaneous collection of samples over large areas. Coupled with boat or ship sampling, remote sensor data allows extrapolation of on-site measurements and provides increased value to traditional sampling approaches. Use of a GIS in combination with on-site sampling and remote sensor data provides an optimal data gathering, storage, and analysis approach for lakes and lacustrine wetland ecosystems.

Water quantity and quality studies have been vastly improved by the use of GIS and re-

## GIS Applications

mote sensor technologies. Research has demonstrated that several water quality variables can be successfully measured, monitored, and modeled using a combination of on-site measurements, remote sensing, and GIS technologies. Water quality variables readily amenable to GIS wetland and lacustrine applications include: surface water temperature, suspended sediments, chlorophyll, dissolved organic carbon, ice characteristics, and others. Evaluations of these variables are particularly good applications of remote sensing and GIS technologies, and other variables may be addressed through surrogate or indirect but correlated variable measurements.

Water quantity issues can be addressed in several ways (Schoolmaster and Marr, 1992). In general, issues of water storage in lakes and fluctuating water levels due to variable runoff and evaporation and transpiration can be evaluated. The capability of near-infrared remote sensing to detect the lake's water edge is particularly valuable.

This capability has been employed from satellite sensors to predict the storage of lakes in shallow basins. These relatively shallow lakes fluctuate greatly in surface area depending on the runoff received. The Great Salt Lake in Utah and Lake Chad in Africa, and other shallow lakes have been evaluated using a combination of remote sensor measurements of lake surface area, topographic/bathymetric maps, and integration of the area and depth relationship using computers or GIS on computers. This approach facilitates the analysis of lake storage in dry or wet times, and allows predictions to be made as to available storage and the potential for flooding of areas adjacent to lakes. Wetlands, both emergent and submergent types, are influenced by the presence and depth of water, and knowledge of their variability is valuable for evaluating risks both ecological and environmental.

## GIS for Water Quality Assessments

Water quality studies have been advanced by the use of a combination of GIS and remote sensor technologies. Remote sensing can supply synoptic measurements of water quality variables or surrogate variables, and GIS can store the data for analyses and modeling from the database.

Initial work in water quality studies has shown that remote sensing and GIS can be very useful in studies of water depth, suspended sediments, concentrations of chlorophyll in phytoplankton, aquatic plants, water temperature, water transport or movements, and ice concentrations and movements. The success of a given application depends on the characteristics of the water body. For example, water depths can be made in water, absent concentrations of suspended sediments. Chlorophyll concentration can be measured from remote sensor data with good results in water with less than 25 mg/l of suspended sediments (Lyon et al., 1988; Ramsey and Jensen, 1996).

Surface-suspended sediment concentration measurements can be useful in monitoring the presence of sediment and loadings, and in monitoring of water movements or transport (Lathrop and Lillesand, 1987; Lyon et al. 1988). Suspended sediment concentrations ranging from approximately 5 mg/l to over 700 mg/l generally exhibit a linear relationship between light upwelling from the water and the surface-suspended sediment concentrations (Lyon et al., 1988; Bhargava and Mariam, 1991; Ritchie and Cooper, 1991; Mertes et al., 1993). The relationship holds true for the red portion, and for the near-infrared portion of the spectrum.

In waters with relatively low concentrations of sediment, the concentration of chlorophyll in phytoplankton can be measured and yields a good agreement with on-site sampling (Hamilton et al., 1993). This is particularly valuable in open ocean measurements, and can be used in large lakes and coastal areas. In particularly clear waters or "optically deep" waters it is necessary to account for the concentration of phytoplankton suspended in the water column, as the penetration and reflectance of light will be influenced by the concentrations throughout the water column that is sensed.

Remote sensing has long been employed for evaluations of water temperature using thermal sensors. The addition of GIS to store and process remote measurements allows the tracking of temperature changes due to phenomena. Great work has been done in coastal, open lake, and ocean area evaluations of surface temperature known as sea surface temperature or SST (Leshkevich et al., 1995).

A major effort has been made in the use of National Oceanic and Atmospheric Administration (NOAA) Advanced Very High Resolution Radiometer (AVHRR) data and GIS for the tracking of coastal and offshore thermal patterns in surface waters. The characteristics of warm and cold core temperature ring structures off the Atlantic coast of the U.S. and other oceans and seas are truly exciting results of these applications of remote sensing and GIS. In lake systems, thermal sensing has been used to study thermal outfalls from natural and human-induced conditions. The combination of technologies of remote sensing to supply data and GIS to process the data in time steps has proven to be a particularly good application, and potentially proven valuable for applications such as evaluating risk to wetlands from changing temperature as a stressor.

Recent work in large lakes has focused on studying phenomena using SST measurements from NOAA AVHRR sensor data. The approximately 1.1 square km resolution is ideal for large-lake studies (Lyon et al., 1988; Bolgrien and Brooks, 1992; Loveland and Ohlen, 1993). Using AVHRR and GIS analysis has been very valuable in the NOAA Coastwatch Program (Miller and DeCampo, 1994; Leshkevich et al., 1995), for study of wind and pressure induced movements of water, presence and movement of ice, the on-set and development of lake stratification, and formation or degradation of the coastal thermal bar. AVHRR data and GIS analysis have also been valuable in studies of large lakes in several areas of the world.

In the presence of clear waters it is possible to use the differential reflectance of light to develop water depth or bathymetry data (Ramsey and Jensen, 1990; Ji et al., 1992). In parts of the Great Lakes and the oceans there exist low concentrations of water colorants, and the waters are often optically deep or the bottom can be sensed from above. In such places, the light-absorbing characteristics of water can be used to sense depth. The combination of visible light remote sensing and GIS technologies and modeling can be used to make water depth maps or bathymetry maps, as well as submergent vegetation or aquatic vegetation maps (Figure 6.6).

A number of experiments conducted by researchers in clear marine or lake waters have yielded high quality results, and the products were employed in additional GIS and other modeling analyses. An interesting project was completed under funding from the Army Corps of Engineers and Ohio Sea Grant, and demonstrated this capability using aircraft remote sensor data, and simple models (Lyon et al., 1992). The work was extended further using complex models (Lyon and Hutchinson, 1995) of light interaction with water.

Dissolved organic carbon can be sensed in lake and marine systems using remote sensing. This capability has been used from AVHRR and Landsat TM data and with databases to study water movement phenomena, such as thermal bar formation on the Great Lakes, and studies of "yellow substance" in ocean waters.

## Lake Water Resource Applications

Some of these applications have been mentioned above to illustrate the use of a particular method or technology. These are mentioned for some of the same reasons to help the reader identify valuable methods that can be adapted for use with either wetland, lake, or coastal water resource applications (Wood et al., 1988; Lunetta et al., 1991; Vieux and Needham, 1992; White et al., 1992; Mitchell, 1995).

The area of riverine and nonpoint source applications has been addressed extensively in the literature, and more work is yet to be completed to truly understand this issue. This attention results from the presence of riverine wetlands, and the prominent role river wetlands play in the lives of biota and humans, and due to the environmental risks associated with waterborne pollutants.

GIS also allows for opportunities to run models or submodels at different places in time and space. This is particularly important for chemical and biological process studies where the materials of interest are changing due to time and location.

Due to the variety of natural variables, wildlife and fishery functions, and human related stressors that influence river and lake characteristics, there has been a great need to employ analysis methods that can integrate multiple variables and evaluate risks to wetlands. GIS has proven valuable to address the complex mix of variables that typically operate in and around lakes. Remote sensing and field measurements have also been valuable in supplying data to complete the picture of lake characteristics and conditions, and to evaluate the influences of humans.

Early work has involved simple evaluations of presence and absence of various resource characteristics and conditions. Many of the capabilities and results available from GIS applications have been described above. Traditional approaches to lake measurements have also proven successful for evaluations, and they can supply valuable data for GIS applications (Hutchinson, 1975; Herdendorf et al., 1981; Linthurst et al., 1986; Messer et al., 1991; Larsen et al., 1995).

Regulatory work has stimulated some early lake management research efforts. The need to manage human development and impacts has required improved methods for monitoring conditions. One important area is the monitoring of compliance with permits (Niedzwiedz and Ganske, 1991). Use of remote sensor and GIS technologies often allows increased productivity from existing people and financial resources in monitoring of permits and other regulatory concerns.

The stimulus for GIS applications in river and lake management has multiple reasons. A particularly important capability is the storage of a variety of data in two-dimensional and three-dimensional formats (Samuels, 1993). GIS are also useful in the analysis of data covering large spatial and temporal scales (Troge, 1995). As is true of many applications, the capability to run models from GIS data sets (Rhodes and Myers, 1993; Corbley, 1994), and develop estimates of status and of trends (McDonnell and MacMillan, 1993) is vital for informed decisions based on numerical analyses of stressors, receptors, exposures, and risks.

The identification and management of riverine or lake or reservoir vegetation is important to lake functions and human uses. Great efforts are made at vegetation or weed control, and methods to identify and track the trend of problem areas are very helpful. A combination of remote sensing measurements and GIS data processing has been found to be a successful approach in lakes and reservoirs of appropriate scale (Jensen et al., 1992, 1993).

## Related Water Resource Applications

There are a number of water resource applications that fall outside the domain of wetland applications per se. Other chapters and other works address these resource applications (Lyon, 2000). In particular, the literature is rich in applications related to rivers (Blick et al., 1987; Argialas et al., 1988), groundwater, point and nonpoint sources of pollution (Lyon, 1987; Kang and Bartholic, 1994), and the like (Goodchild et al., 1993; Kovar and Nachtnebel, 1993; Sample, 1994; Lyon and McCarthy, 1995; Lyon, 1998).

Riverine and watershed applications have witnessed many high quality applications of GIS. GIS are particularly suited to the requirements of these studies (Jenson and Dominque, 1988; Maidment, 1993a, 1993b; Moore et al., 1993; Nyerges, 1993; Sample, 1994; Thenkabail et al., 2000). GIS allow the user to define and integrate measurements within Hydrologic Resource Units (HRU) of a variety of sizes. With GIS it is possible to route the movement of water and waterborne materials from the subwatershed to watershed to the rivers and other downstream entities (Fast and Rajala, 1995; Lyon, 1999).

Most of the methods used in riverine (Ervin, 1992; Horn and Grayman, 1992; Olson, 1992; Clifford et al., 1995; Green et al., 1995; Richards et al., 1995; Senay et al., 2000) and groundwater (Pickus and Hewitt, 1992) studies can be adapted to studies of wetland, lake, and coastal ecosystems. The issues of protecting water quality and nonpoint sources of pollution affects all these resources (Griner, 1993; Cahill et al., 1995). Nonpoint source and fate studies can often supply insight as to how a particular problem can be addressed for non-riverine resources. Nonpoint source and GIS applications to wetlands have been evaluated and are particular enlightening (Poiani and Bedford, 1995).

# 7 Evaluations of Accuracy

## Assessments of Data Quality

As is true of pioneering efforts, many priorities must be addressed and users learn along the way. When faced with huge data collection efforts, achieving results is vital. Naturally, work is conducted with high levels of quality. Now, it is even more important to document the quality of data and assure their collection using standard procedures. The procedures to accomplish data quality assurance (QA) and data quality control (QC) in data management collection are well known in laboratory science.

These QA/QC concepts have been introduced into GIS, remote sensor, and landscape characterization efforts. Previously, the actual writing of standards and actual assessments of control were informal and performed occasionally. This was due to cost considerations and due to the hardships associated with developing and implementing new technologies such as GIS and regional-scale measurements from remote sensors. In many cases the QA/QC issues had been addressed but may not have been formally established and documented.

Many groups now require the use of QA/QC procedures and assessments of accuracy. An example is the U.S. Environmental Protection Agency, where such procedures are Agency-wide in their practice.

## Quality Assurance/Quality Control (QA/QC) Issues and Approaches

QA/QC issues are manifold in GIS and remote sensor technologies and their application. They involve the control of work using performance standards and reporting procedures that supply heritage for measurements. They also involve evaluations of accuracy and precision downstream in the effort, that may include assurance that procedures were followed and documented (Thapa and Bossler, 1992). Included are independent assessments of the quality of results by comparison with other, independent data, so as to assess the success of a given product against other independent benchmark data sets.

QA/QC methods in GIS and remote sensing can be categorized into several groups. These categories may follow the flow of data collection to analysis, and to results and products.

An early and frequent effort is the documentation of performance of each step. The documentation of the performance of QC procedures and testing of products is vital. This is due to the fact that applications such as these usually require a number of procedures conducted over large areas, and using many researchers. This inherent level of variability in resources, and in people and equipment measuring resources, argues for a high level of control and assurance. Otherwise, there can be diminished confidence in the results and products.

An important effort is the use of agreed-upon techniques codified by project standards allowing for common performance of procedures. Documentation of the performance of procedures lends a paper trail to the work and a heritage of data processing efforts.

QC standards should be developed to ensure the collection and processing of GIS data in a uniform and correct manner. At the onset of a project or application, the research team needs to generate standards or procedures of performance for each step of the project (Lunetta et al., 1993). These may be quantitative in characteristic, or qualitative if a quantitative description is not appropriate. These may take the form of steps in procedures to be followed by researchers, or performance standards as to the outcome of a certain step. While time-consuming to develop, such QC standards can be invaluable.

## Assessments of Data Accuracy

A powerful tool in the determination of the accuracy of a product is a comparison with an independent data set that measures the same variables of interest (Congalton, 1991). In an application there is usually a resulting product. Oftentimes this is a map, image, or matrix of resource classes. The product is derived from considerable effort and is the yield from many analyses and procedural steps. As such, it is difficult to perform some internal, statistical analysis and derive a meaningful result that measures that total. The product must be subjected to an assessment of accuracy or quality assurance test.

These remote sensor or GIS data are similar in composition to the application product, yet are available from separate sources. A project may develop a GIS product that identifies the location of general wetland types and models the predicted functions of the types. This example product may be developed for a regional-sized area. To determine the accuracy of each general wetland type and function for the entire area would be cost-prohibitive, and it would defeat the intent of the application, as a 100% assessment of accuracy would be redundant. In this example, a subset of the entire region can be survey sampled and the accuracy of the application product and its general wetland types and functions can be compared to the actual subset that was sampled. This is a great saving in time and cost, and can be highly accurate.

Survey sampling procedures such as this can also be performed to yield a given confidence level in the result. That is, the QA/QC plan for the project can state that a given type-class must be accurate at a given level for a given confidence interval or range. The size of the independent survey sample can be calculated so as to be sure to select enough samples in the region-sized area to adequately evaluate the accuracy of the application product.

## Data Analysis

To further enhance the value of the application product, it is desirable to make calculations of the numbers and types of type-class errors identified, as compared to the independent-type classes and functions generated in the survey sample (ASPRS, 1990). This allows the user to determine the accuracy of a given type-class, and to determine the identity of erroneous determinations. This information is powerful because it addresses errors of commission and omission (Congalton, 1991), and provides the identities of the errors encountered. This information often allows the user or later users to make decisions as to the appropriate form of the product, and whether it is suitable for their given application (Story and Congalton, 1986).

Once the survey sample is completed, the individual type-class accuracies can be calculated. This is a valuable assessment of accuracy. The best way to store all this accuracy assessment data is a $2 \times 2$ matrix called the confusion matrix or error matrix.

The results from the application product are provided on one axis of the matrix and the results of the independent data set or survey sample are stored on the other axis. The diagonal elements of the matrix store the number of correct classes, and the sum of the diagonals is the number of samples that were correctly identified by the product as determined from the independent data set. The off-diagonal elements store the incorrect classes and the numbers of incorrect samples found.

These matrices allow the user to identify the accuracy of the product classes. It also allows the user to identify for a given product type-class which other classes were incorrectly identified, and the type of incorrect identifications (Congalton and Schallert, 1992).

The error matrix becomes a powerful tool for the producer of the product. The producer can judge the accuracy of the product and identify where the inaccuracies occurred (Bolstad and Smith, 1995). The later user of the product can identify the accuracy of product type-classes and also see the sort of inaccuracies involved. Oftentimes, the inaccuracies involved may not interfere with the work of the later users. The knowledge supplied by the error matrices is truly a great help to the user, and it supplies QA/QC information to both the producer and the user.

# *Bibliography*

Ackleson, S. and V. Klemas, 1987. Remote sensing of submerged aquatic vegetation in lower Chesapeake Bay: A comparison of Landsat MSS to TM imagery. *Remote Sensing of Environment* 22:235–248.

Adamus, P., L. Stockwell, E. Clairain, M. Morrow, L. Rozas, and R. Smith, 1991. Wetland Evaluation Technique (WET). Technical Report WRP-DE-2, Waterways Experiment System, U.S. Army Corps of Engineers.

Anderson, G. and D. Hanson, 1992. Evaluating hand-held radiometer derived vegetation indices for estimating above ground biomass. *Geocarto International* 1:71–77.

Anderson, J., E. Hardy, J. Roach, and R. Witmer, 1976. A Land Use Classification System for Use With Remote-Sensor Data. U.S. Department of Interior, U.S. Geological Survey Professional Paper 964, Washington, D.C.

Argialas, D., J. Lyon, and O. Mintzer, 1988. Quantitative description and classification of drainage patterns. *Photogrammetric Engineering and Remote Sensing* 54:505–509.

Arnold, J., 1992. Using GIS and remote sensing to identify and monitor oilfield waste pits in Louisiana's coastal zone. *Proceedings of the First Thematic Conference on Remote Sensing of Marine and Coastal Environments*, ERIM, Ann Arbor, MI, pp.707–718.

ASPRS, 1990. ASPRS accuracy standards for large-scale maps. *Photogrammetric Engineering and Remote Sensing* 56:1068–1070.

Augenstein, E., D. Stow, and A. Hope, 1991. Evaluation of SPOT HRV-XS data for kelp resource inventories. *Photogrammetric Engineering and Remote Sensing* 57:501–509.

Bagheri, S. and M. Stein, 1992. Monitoring of Judson Rarital estuary using multispectral video data. *International Journal of Remote Sensing* 13:965–969.

Balogh, M., L. Fisher, A. Bailey, and R. Lunetta, 1992. Application of GPS and aircraft MSS data to a Puget Sound intertidal habitat study. *Proceedings of the First Thematic Conference on Remote Sensing of Marine and Coastal Environments*, ERIM, Ann Arbor, MI, pp.695-696.

Banner, A. and T. Lynham, 1981. Multitemporal analysis of Landsat data for forest cutover mapping—A trial of two procedures. *Proceedings of the Seventh Canadian Symposium on Remote Sensing*, pp. 233-240.

Bertram, T. and A. Cook, 1993. Satellite imagery and GPS-aided ecology. *GPS World*, October, pp.48-53.

Bhargava, D. and D. Mariam, 1991. Effects of suspended particle size and concentration on reflectance measurements. *Photogrammetric Engineering and Remote Sensing* 57:519-529.

Blick, D., J. Messer, D. Landers, and W. Overton, 1987. Statistical basis for the design and interpretation of the National Stream Survey, phase I:Lakes and streams. *Lake and Reservoir Management* 3:470-475.

Bolgrien, D. and A. Brooks, 1992. Analysis of thermal features of Lake Michigan from AVHRR satellite images. *Journal of Great Lakes Research* 18:259-266.

Bolstad, P. and J. Smith, 1995. Errors in GIS, assessing spatial data accuracy. In Lyon, J. and J. McCarthy, 1995, pp. 301-312.

Breininger, D., M. Provancha, and R. Smith, 1991. Mapping Florida Scrub Jay habitat for purposes of land-use management. *Photogrammetric Engineering and Remote Sensing* 57:1467-1474.

Brewster, A. and H. Monday, 1994. Airborne imagery used in wetlands restoration. *Earth Observation Magazine*, August, pp.57-59.

Brinson, M., 1993. A Hydrogeomorphic Classification for Wetlands. Waterways Experimental Station, U.S. Army Corps of Engineers, WRP-DE-4, Washington, D.C.

Brondizio, E., E. Moran, P. Mausel, and Y. Wu, 1996. Land cover in the Amazon estuary: Linking of the Thematic Mapper with botanical and historical data. *Photogrammetric Engineering and Remote Sensing* 62:921-929.

# Bibliography

Browder, J., L. May, A. Rosenthal, J. Gosselink, and R. Baumann, 1989. Modeling future trends in wetland loss and Brown Shrimp production in Louisiana using Thematic Mapper imagery. *Remote Sensing of Environment* 28:45-59.

Brown, D., C. Lowe, and C. Pase, 1979. A digitized classification system for the biotic communities of North America, with community (series) and association examples for the Southwest. *Journal of the Arizona-Nevada Academy of Science* 14:1-16.

Bukata, R., J. Bruton, J. Jerome and W. Haras, 1987. A Mathematical Description of the Effects of Prolonged Water Level Fluctuations on the Areal Extent of Marshland. Report RRB-87-02, Canada Centre for Inland Waters, Burlington, Ontario, Canada.

Butera, K., 1983. Remote sensing of wetlands. *IEEE Transactions on Geoscience and Remote Sensing* GE-21:383-392.

Byrne, G., P. Crapper, and K. Mayo, 1980. Monitoring landcover change by principal compnent analysis of multitemporal Landsat data. *Remote Sensing of the Environment* 10:175-184.

Cahill, T., W. Horner, and J. McGuire, 1995. GIS watershed applications in the analysis of non-point source pollution. In USEPA, 1995, pp.110-130.

Carter, V., 1982. Applications of remote sensing to wetlands. In Johannsen, C. and J. Sanders, Eds., *Remote Sensing for Resource Management*, Soil Conservation Society of American, Ankeny, IA, pp.284-300.

Carter, V., 1990. Importance of Hydrologic Data for Interpreting Wetland Maps and Assessing Wetland Loss and Mitigation. Federal Coastal Wetland Mapping Programs, Biology Report 90 (18), U.S. Fish and Wildlife Service, Washington, D.C., pp.79-85.

Channell, W., 1989. Selected applications of the Ohio Capability Analysis Program. *Proceedings of the 1989 Fall Convention of ACSM/ASPRS*, Cleveland, OH, pp.23-37.

Christel-Rose, L. and F. Scarpace, 1991. Monitoring the spatial distribution of aquatic macrophytes: A look at plant associations in Allequash lake, Wisconsin. *Proceedings of the Annual Convention of ACSM/ASPRS*, Vol. 4, GIS, pp.21-30.

Clark, J., 1996. *Coastal Zone Management Handbook*. Lewis Publishers, Boca Raton, FL.

Clifford, J., W. Wheaton, and R. Curry, 1995. EPA's reach-indexing project—using GIS to improve water quality assessment. In USEPA, 1995, pp.242-252.

Cochran, W., 1977. *Sampling Techniques*. John Wiley and Sons, New York, NY.

Colwell, J. and F. Weber, 1981. Forest Change Detection. *Proceedings of the 15th International Symposium on Remote Sensing of Environment*, Environmental Research Institute of Michigan, Ann Arbor, MI, pp.65-99.

Congalton, R., 1991. A review of assessing the accuracy of classifications of remotely sensed data. *Remote Sensing of Environment* 37:35-46.

Congalton, R., 1996. A Quantitative Comparison of Change Detection Algorithms for Monitoring Eelgrass from Remotely Sensed Data. Unpublished manuscript, University of New Hampshire, Durham, NH.

Congalton, R. and K. Green, 1998. *Accuracy Assessment of Remotely Sensor Data: Principles and Practices*. CRC/Lewis Publishers, Boca Raton, FL.

Congalton, R. and D. Schallert, 1992. Exploring the Effects of Vector to Raster and Raster to Vector Conversion. U.S. Environmental Protection Agency, EPA/600/R-92/166, Las Vegas, NV.

Corbley, K., 1994. Lake Michigan GIS models: Possible solutions to area ozone problem. *Earth Observation Magazine,* July, pp.16-18.

Cowardin, L., V. Carter, F. Golet, and E. LaRoe, 1979. Classification of Wetlands and Deepwater Habitats of the United States. U.S. Department of Interior, U.S. Fish and Wildlife Service, Report No. FWS/OBS-79/31, Washington, D.C.

Derring, D. and R. Haas, 1980. Using Landsat Digital Data for Estimating Green Biomass. NASA Technical Memorandum #80727.

Deysher, L., R. Ayers, R. Grove, and A. Jahn, 1995. GIS analysis of kelp canopy remote sensing data. *Proceedings of the Third Thematic Conference on Remote Sensing of Marine and Coastal Environments*, ERIM, Ann Arbor, MI, Vol. 1, pp.700-708.

Dobson, J. and E. Bright, 1993. Large-area change analysis: The Coastwatch Change Analysis Project (CCAP). *Proceedings of the 12th Pecora Symposium*, Sioux Falls, SD, pp.73-81.

Donoghue, D., D. Thomas, and Y. Zong, 1994. Mapping and monitoring the intertidal zone of the east coast of England using remote sensing techniques and a coastal monitoring GIS. *Marine Technology Society Journal* 28:19-29.

# Bibliography

Douglas, W., 1995. *Environmental GIS, Applications to Industrial Facilities*. Lewis/CRC Publishers, Boca Raton, FL. 144 pp.

Eidenshink, J. and R. Haas, 1992. Analyzing vegetation dynamics of land system with satellite data. *Geocarto International* 1:53-61.

Elevand, M., C. Tao, and R. VanZuidam, 1995. Towards a decision support system for coastal zone management by applying morphodynamic modeling with remote sensing data inputs in a 4-D GIS environment. *Proceedings of the Third Thematic Conference on Remote Sensing of Marine and Coastal Environments*, ERIM, Ann Arbor, MI, Vol.1, pp.256-265.

Elzinga, C. and A. Evenden, 1997. Vegetation Monitoring: An Annotated Bibliography. General Technical Report INT-GTR-352, Fort Collins, CO, U.S. Department of Agriculture, Forest Service.

Ervin, S., 1992. Integrating visual and environmental analyses in site planning and design. *GIS World*, July, pp.26-30.

Falkner, E., 1994. *Aerial Mapping Methods and Applications*. Lewis/CRC Publishers, Boca Raton, FL.

Fast, M. and T. Rajala, 1995. Decision support system for multiobjective riparian/wetland corridor planning. In USEPA, 1995, pp.213-217.

Federal Geographic Data Committee (FGDC), 1992. Application of Satellite Data for Mapping and Monitoring Wetlands. Wetlands Subcommittee, Washington, D.C.

Federal Interagency Committee for Wetlands Delineation, 1989. Federal Manual for Identifying and Delineating Jurisdictional Wetlands. January 10, U.S. Government Printing Office, Washington, D.C.

Fedra, K., 1993. Models, GIS, and expert systems: Integrated water resources models. *Proceedings of the HydroGIS'93,* International Associate of Hydrological Sciences Publication No.211.

Field, D., A. Reyer, C. Alexander, B. Shearer, and P. Genovese, 1990. NOAA National Coastal Wetlands Inventory. Federal Coastal Wetland Mapping Programs, Biology Report 90 (18), U.S. Fish and Wildlife Service, Washington, D.C., pp.39-49.

Friel, C., W. Sargent, and C. Westlake, 1995. Can GIS help save Florida bay? *GIS World,* pp.40-45.

Fung, T. and E. LeDrew, 1987. Application of principal component analysis to change detection. *Photogrammetric Engineering and Remote Sensing* 53:1649-1658.

Gahegan, M. and J. Flack, 1996. A model to support the integration of image understanding techniques within a GIS. *Photogrammetric Engineering and Remote Sensing* 62:483-490.

Garbrecht, J. and L. Martz, 1993. Network and subwatershed parameters extracted from digital elevation models: The Bills creek experience. *Water Resources Bulletin* 29:909-916.

Good, R., D. Whigham, and R. Simpson, 1978. *Freshwater Wetlands: Ecological Processes and Management Potential.* Academic Press, New York, NY.

Goodchild, M., B. Parks, and L. Steyaert, 1993. *Environmental Modeling with GIS.* Oxford University Press, Oxford, United Kingdom.

Gosselink, J., L. Lee, and T. Muir, 1990. *Ecological Processes and Cumulative Impacts.* Lewis Publishers, Boca Raton, FL.

Graham, L., 1993. Airborne video for near real-time vegetation mapping. *Journal of Forestry* 91:28-32.

Green, K, S. Bernath, L. Lackey, M. Brunego, and S. Smith, 1995. Analyzing the cumulative effects of forest practices: Where do we start? In Lyon, J. and J. McCarthy, 1995.

Griffin, C., 1995. Data quality issues affecting GIS use for environmental problem-solving. In USEPA, 1995, pp.15-30.

Griffin, W. and F. Sedgwick, 1995. Digital orthophotos: The basis for cost sharing. *Earth Observation Magazine* pp.40-43.

Griner, A., 1993. Development of a water supply protection model in a GIS. *Water Resources Bulletin* 29:965-971.

Gross, M., M. Hardsky, V. Klemas, and P. Wolf, 1987. Quantification of biomass of the marsh grass *Spartina Alterniflora loisel* using Landsat Thematic Mapper imagery. *Photogrammetric Engineering and Remote Sensing* 53:1577-1583.

Hamilton, M., C. Davis, W. Rhea, S. Pilorz, and K. Carder, 1993. Estimating chlorophyll content and bathymetry of Lake Tahoe using AVIRIS data. *Remote Sensing of Environment* 44:217-230.

# Bibliography

Hammer, D., 1989. *Constructed Wetlands for Wastewater Treatment: Municipal, Industrial and Agricultural*. Lewis Publishers, Chelsea, MI.

Hammer, D., 1992. *Creating Freshwater Wetlands*. Lewis Publishers, Chelsea, MI.

Harris, A., 1994. Time series remote sensing of a climatically sensitive lake. *Remote Sensing of Environment* 50:83-94.

Hastings, D. and L. Di, 1994a. Modeling of global change phenomena with GIS using the global change data base. I: Modeling with GIS. *Remote Sensing of Environment* 49:1-12.

Hastings, D. and L. Di, 1994b. Modeling of global change phenomena with GIS using the global change data base. I: Prototype synthesis of the AVHRR-based vegetation index from terrestrial data. *Remote Sensing of Environment* 49:13-24.

Heinen, J. and J. Lyon, 1989. The effects of changing weighting factors on wildlife habitat index values: A sensitivity analysis. *Photogrammetric Engineering and Remote Sensing* 55:1445-1447.

Herdendorf, C., S. Hartley, and M. Barnes, 1981. Fish and Wildlife Resources of the Great Lakes Coastal Wetlands Within the United States. FWS/OBS-81/02-VI, U.S. Fish and Wildlife Service, Washington, D.C.

Herr, A. and L. Queen, 1993. Crane habitat evaluation using GIS and remote sensing. *Photogrammetric Engineering and Remote Sensing* 59:1531-1538.

Hewitt, M., 1990. Synoptic inventory of riparian ecosystems: The utility of Landsat Thematic Mapper data. *Forest Ecology and Management* 34:605-620.

Hook, D., B. Davis, J. Scott, J. Struble, C. Bunton, and E. Nelson, 1995. Locating delineated wetland boundaries in coastal South Carolina using global positioning systems. *Wetlands* 15:31-36.

Horn, C. and W. Grayman, 1992. Water-quality modeling with EPA reach file system. *Journal of Water Resources Planning and Management* 119:262-274.

Hruby, T., W. Cesanek, and K. Miller, 1995. Estimating relative wetland values for regional planning. *Wetlands* 15:93-107.

Hussin, Y. and R. Hoffer, 1989. Relationship between multipolarized radar backscatter and Slash Pine stand parameters. *Proceedings of the 1989 Fall Convention of ACSM/ASPRS*, Cleveland, OH, pp.184-196.

Hutchinson, G., 1975. *A Treatise on Limnology, Volume III, Limnological Botany*. J. Wiley, New York, NY.

Hutchinson, W., 1989. Application of a radiometric model and remote sensor data for evaluation of water depths. Master's thesis, Department of Civil Engineering, Ohio State University, Columbus, OH.

Jakubauskas, M., J. Whistler, M. Dillworth, and E. Martinko, 1992. Classifying remotely sensed data for use in an agricultural nonpoint source pollution model. *Journal of Soil and Water Conservation* 47:179-183.

Jaworski, E., C. Raphael, P. Mansfield, and B. Williamson, 1979. Impact of Great Lakes Water Level Fluctuations on Coastal Wetlands. Institute for Water Research, Michigan State University, East Lansing, MI.

Jensen, J., 1996. *Introductory Digital Image Processing: A Remote Sensing Perspective*. Prentice Hall, Englewood Cliffs, NJ.

Jensen, J., S. Narumaiani, O. Weatherbee, K. Morris, and H. Mackey, 1992. Predictive modeling of cattail and waterlily distribution in a South Carolina reservoir. *Photogrammetric Engineering and Remote Sensing* 58:1561-1568.

Jensen, J., S. Narumaiani, O. Weatherbee, and H. Mackey, 1993. Measurement of seasonal and yearly cattail and waterlily changes using multidate SPOT panchromatic data. *Photogrammetric Engineering and Remote Sensing* 59:519-525.

Jenson, S. and J. Domingue, 1988. Extracting topographic structure from digital elevation data for geographic information system analysis. *Photogrammetric Engineering and Remote Sensing* 54:1593-1600.

Ji., W., D. Civco, and W. Kennard, 1992. Satellite remote bathymetry: A new mechanism for modeling. *Photogrammetric Engineering and Remote Sensing* 58:545-549.

Ji, W. and L. Mitchell, 1995. Analytical model-based decision support GIS for wetland resource management. In Lyon and McCarthy, 1995, pp.31-47.

# Bibliography

Johnston, C., N. Detenbeck, J. Bonde, and G. Niemi, 1988. Geographic Information Systems for cumulative impact assessment. *Photogrammetric Engineering and Remote Sensing* 54:1609-1615.

Johnston, C., N. Detenbeck, and G. Niemi, 1990. The cumulative effect of wetlands on streamwater quality and quantity, a landscape approach. *Biogeochemistry* 10:105-141.

Johnston, C., 1994. Cumulative impacts to wetlands. *Wetlands* 14:49-55.

Johnston, J. and L. Handley, 1990. Coastal Mapping Programs at the U.S. Fish and Wildlife Service's National Wetlands Research Center. Federal Coastal Wetland Mapping Programs, Biology Report 90 (18), U.S. Fish and Wildlife Service, Washington, D.C., pp.105-109.

Kang, Y. and J. Bartholic, 1994. A GIS-based agricultural nonpoint source pollution management system at the watershed level. *Proceedings of the Annual ACSM/ASPRS Convention*, Reno, NV, pp.281-289.

Kang, Y., T. Zahniser, L. Wolfson, and J. Bartholic, 1994. WIMS: A prototype wetlands information management system for facilitating wetland decision making. *Proceedings of the Annual ACSM/ASPRS Convention*, Reno, NV, pp.290-300.

Kennedy, M., 1996. *The Global Positioning System and GIS*. Ann Arbor Press, Chelsea, MI.

Klemas, V., R. Gantt, H. Hassan, N. Patience, and O. Weatherbee, 1995. Environmental Information Systems for Coastal Zone Management. Environmental Department, the World Bank.

Kovar, K. and H. Nachtnebel, 1993. Application of Geographic Information Systems in Hydrology and Water Resources Management. International Association of Hydrological Scientists, Publication No. 211, Wallingford, United Kingdom.

Kusler, J. and M. Kentula, 1990. Wetland Creation and Restoration. Association of State Wetland Managers, Berne, NY.

Lampman, J., 1993. Bibliography of Remote Sensing Techniques Used in Wetland Research. National Technical Information Service, Springfield, VA.

Larsen, D., N. Urquhart, and D. Kugler, 1995. Regional scale trend monitoring of indicators of trophic condition of lakes. *Water Resources Bulletin* 31:117-140.

Lathrop, R. and T. Lillesand, 1987. Calibration of Thematic Mapper thermal data for water surface temperature mapping: A case study on the Great Lakes. *Remote Sensing of Environment* 22:297-307.

Lee, C. and S. Marsh, 1995. The use of archival Landsat MSS and ancillary data in a GIS environment to map historical change in an urban riparian habitat. *Photogrammetric Engineering and Remote Sensing* 61:999-1008.

Lee, J., R. Park, and P. Mausel, 1992. Application of geoprocessing and simulation modeling to estimate impacts of sea level rise on the northeast coast of Florida. *Photogrammetric Engineering and Remote Sensing* 58:1579-1586.

Lee, K. and R. Lunetta, 1990. Watershed Characterization Using Landsat Thematic Mapper (TM) Satellite Imagery, Lake Pend Oreille, Idaho. Report TS-AMD-90C10, U.S. Environmental Protection Agency, Las Vegas, NV.

Lee, K. and R. Lunetta, 1993. Great Lakes Ecological Process Pilot (GLEPP) Green Bay, Wisconsin. Interim report, U.S. Environmental Protection Agency, Las Vegas, NV.

Lee, K. and R. Lunetta, 1995. Wetland detection methods investigation. In Lyon and McCarthy, 1995, pp.248-264.

Leshkevich, G., D. Schwab, and G. Muhr, 1995. Satellite environmental monitoring of the Great Lakes: Great Lakes coastwatch program update. *Proceedings of the Third Conference of Remote Sensing for Marine and Coastal Environments*, Seattle, WA, I-116-127.

Linthurst, R., D. Landers, J. Eilers, D. Brakke, W. Overton, E. Meier, and R. Crowe, 1986. Eastern Lake Survey-Phase I, Characteristics of Lakes in the Eastern United States. Technical Report EPA/600/4-86/007a, U.S. Environmental Protection Agency, Las Vegas, NV.

Lo, C. and L. Watson, 1994. Okefenokee swamp vegetation mapping with Landsat Thematic Mapper data: An evaluation. *Proceedings of the Annual ACSM/ASPRS Convention*, Reno, NV, pp.365-374.

Loveland, T. and D. Ohlen, 1993. Experimental AVHRR land data sets for environmental monitoring and modeling. In Goodchild et al., 1993, pp.379-385.

Lunctta, R. and C. Elvidge, 1998. *Remote Sensing Change Detection*, Ann Arbor Press, Chelsea, MI. 318 pp.

# Bibliography

Lunetta, R., R. Congalton, L. Fenstermaker, J. Jensen, K. McGwire, and L. Tinney, 1991. Remote sensing and Geographic Information System data integration: Error sources and research issues. *Photogrammetric Engineering and Remote Sensing* 57:677-688.

Lunetta, R., J. Lyon, D. Worthy, J. Sturdevant, J. Dwyer, D. Yuan, C. Elvidge, and L. Fenstermaker, 1993. North American Landscape Characterization (NALC), Landsat Pathfinder Technical Plan. U.S. Environmental Protection Agency, EPA 600/X-93/009, Las Vegas, NV.

Lunetta, R., J. Lyon, C. Elvidge, and B. Guindon, 1998. North American Landscape Characterization: Dataset development and data fusion issues. *Photogrammetric Engineering and Remote Sensing* 64:821-829.

Lunetta, R. and M. Balogh, 1999. Application of multi-temporal Landsat 5 TM imagery to wetland identification. *Photogrammetric Engineering and Remote Sensing* 65:1303-1310.

Lyon, J., 1978. An analysis of vegetation communities in the Lower Columbia River Basin. *Proceedings of the Pecora Symposium on Applications of Remote Sensing to Wildlife Management*, Sioux Falls, SD, pp.321-327.

Lyon, J., 1979. Remote sensing of coastal wetlands and habitat quality of the St. Clair Flats, Michigan. *Proceedings of the 13$^{th}$ International Symposium on Remote Sensing of Environment*, Ann Arbor, MI, pp. 1117-1129.

Lyon, J., 1980. Data sources for analyses of Great Lakes Wetlands. *Proceedings of the Annual Meeting of the American Society for Photogrammetry*, St. Louis, MO, pp. 512-525.

Lyon, J., 1981. The influence of Lake Michigan water levels on wetland soils and distribution of plants in the Straits of Mackinac, Michigan. Doctoral dissertation, University of Michigan, Ann Arbor, MI.

Lyon, J., 1983. Landsat-derived land cover classification for locating potential Kestrel nesting habitat. *Photogrammetric Engineering and Remote Sensing* 49:245-250.

Lyon, J., 1987. Maps, aerial photographs and remote sensor data for practical evaluations of hazardous waste sites. *Photogrammetric Engineering and Remote Sensing* 53:515-519.

Lyon, J., 1993. *Practical Handbook for Wetland Identification and Delineation*. CRC/Lewis, Boca Raton, FL.

Lyon, J., 1995. Wetlands: How to avoid getting soaked. *Professional Surveyor* 15:16-18.

Lyon, J., 2000. *GIS for Water Resources and Watershed Management*. Ann Arbor Press, Chelsea, MI.

Lyon, J. and K. Adkins, 1995. Use of a GIS for wetland identification, the St. Clair Flats, Michigan. In Lyon and McCarthy, 1995, pp.49-58.

Lyon, J. and R. Drobney, 1984. Lake level effects as measured from aerial photos. *Journal of Surveying Engineering* 110:103-111.

Lyon, J. and R. Greene, 1992. Lake Erie water level effects on wetlands as measured from aerial photographs. *Photogrammetric Engineering and Remote Sensing* 58:1355-1360.

Lyon, J. and W. Hutchinson, 1995. Application of a radiometric model for evaluation of water depths and verification of results with airborne scanner data. *Photogrammetric Engineering and Remote Sensing* 61:161-166.

Lyon, J. and J. McCarthy, 1981. Seasat radar imagery for detection of coastal wetlands. *Proceedings of the 15th International Symposium on Remote Sensing of Environment*, Ann Arbor, MI, pp.1475-1485.

Lyon, J. and J. McCarthy, 1995. *Wetland and Environmental Applications of GIS*. Lewis Publishers, Boca Raton, FL.

Lyon, J. and C. Olson, 1983. Inventory of Coastal Wetlands. Michigan Sea Grant Program Publication, University of Michigan, Ann Arbor, MI.

Lyon, J., R. Drobney, and C. Olson, 1986. Effects of Lake Michigan water levels on wetland soil chemistry and distribution of plants in the Straits of Mackinac. *Journal of Great Lakes Research* 12:175-183.

Lyon, J., E. Falkner, and W. Bergen, 1995. Cost estimating photogrammetric and aerial photography services. *Journal of Surveying Engineering* 121:63-86.

Lyon, J., R. Lunetta, and D. Williams, 1992. Airborne multispectral scanner data for evaluation of bottom types and water depths of the St. Marys River, Michigan. *Photogrammetric Engineering and Remote Sensing* 58:951-956.

Lyon, J., J. McCarthy, and J. Heinen, 1986. Video digitization of aerial photographs for measurement of wind erosion damage on converted rangeland. *Photogrammetric Engineering and Remote Sensing* 52:373-377.

# Bibliography

Lyon, J., D. Williams, and K. Flanigan, 1994. Effects of commercial vessel passage in narrow channels with and without ice cover. *Journal of Cold Regions Engineering* 8:47-64.

Lyon, J., K. Bedford, J. Chien-Ching, D. Lee, and D. Mark, 1988. Suspended sediment concentrations as measured from multidate Landsat and AVHRR data. *Remote Sensing of Environment* 25:107-115.

Lyon, J., J. Heinen, R. Mead, and N. Roller, 1987. Spatial data for modeling wildlife habitat. *Journal of Surveying Engineering* 113:88-100.

Lyon, J., D. Yuan, R. Lunetta, and C. Elvidge, 1998. A change detection experiment using vegetation indices. *Photogrammetric Engineering and Remote Sensing* 64:143-150.

Lytle, D., 1993. Digital soils database for the United States. In Goodchild et al., 1993, pp.386-391.

Maidment, D., 1993a. GIS and hydrologic modeling. In Goodchild et al., 1993, pp.147-167.

Maidment, D., 1993b. Developing a spatially distributed unit hydrography by using GIS. In Kovar and Nachtnebel, 1993, pp.181-192.

Marble, A., 1992. *A Guide to Wetland Functional Design*. Lewis Publishers, Chelsea, MI.

Martz, L. and J. Garbrecht, 1993. Automated extraction of drainage network and watershed data from digital elevation models. *Water Resources Bulletin* 29:901-908.

McDonnell, R. and W. MacMillan, 1993. A GIS-based hierarchical simulation model for assessing the impacts of large dam projects. In Kovar and Nachtnebel, 1993, pp.409-416.

McKee, P., T. Batterson, T. Dahl, V. Glooschenko, E. Jaworski, J. Pearce, C. Raphael, T. Whillans, and E. LaRoe, 1988. Great Lakes aquatic habitat classification based on wetland classification systems. In the Development of an Aquatic Habitat Classification System for Lakes, pp.59-72.

Mertes, L., M. Smith, and J. Adams, 1993. Estimating suspended sediment concentrations in surface waters of the Amazon River wetlands from Landsat images. *Remote Sensing of Environment* 43:281-301.

Messer, J., R. Linthurst, and W. Overton, 1991. An EPA program for monitoring ecological status and trends. *Environmental Monitoring and Assessment* 17:67-78.

Miller, R. and J. DeCampo, 1994. CCOAST: A PC-based program from the analysis of coastal processes using NOAA CoastWatch data. *Photogrammetric Engineering and Remote Sensing* 60:155-159.

Mitchell, J., 1995. GIS uncertainty and policy: Where do we draw the 25-inch line? In USEPA, 1995, pp.3-14.

Mitsch, W. and J. Gosselink, 1993. *Wetlands*. Van Nostrand Reinhold, New York, NY.

Moik, J., 1980. Digital Processing of Remotely Sensed Images. National Aeronautics and Space Administration, NASA SP-431, Washington, D.C.

Moore, I., A. Turner, J. Wilson, S. Jenson and L. Band, 1993. GIS and land surface-subsurface process modeling. In Goodchild, 1993, pp.196-230.

Mulchoney, D., D. Grossman, and R. Solomon, 1991. Rapid ecological assessment for conservation planning. *Proceedings of the Annual Convention of ACSM/ASPRS*, Vol. 4, GIS, pp.141-145.

Mynar, F. and R. Lunetta, 1990. Classification of Puget Sound Nearshore Habitats Using Aircraft Multispectral Scanner Imagery. Report TS-AMD-90C11, U.S. Environmental Protection Agency, Las Vegas, NV.

National Research Council, 1995. *Wetlands: Characteristics and Boundaries*. National Academy Press, Washington, D.C.

Natural Resource Conservation Service, 1997. Hydrology tools for wetland determination. Chapter 19 in *Engineering Field Handbook*, U.S. Department of Agriculture, Washington, D.C.

Nawrocki, T., 1995. Design of GIS analysis to compare wetland impacts on runoff in upstream basins of the Mississippi and Volga rivers. In USEPA, 1995, pp.218-232.

Nelson, R., 1983. Detecting forest canopy change due to insect activity using Landsat MSS. *Photogrammetric Engineering and Remote Sensing* 49:1303-1314.

Niedzwiedz, W. and L. Ganske, 1991. Assessing lakeshore permit compliance using low altitude oblique 35-mm aerial photography. *Photogrammetric Engineering and Remote Sensing* 57:511-518.

Niedzwiedz, W. and S. Batie, 1984. An assessment of urban development into coastal wetlands using historical aerial photography: A case study. *Environmental Management* 8:205-214.

Norton, D. and E. Slonecker, 1990. The ecological geography of EMAP. *Geo Info Systems* 1:33-43.

Novitzki, R., 1979. Hydrological characteristics of Wisconsin's wetlands and their influence on floods, stream flow, and sediment. In Greenson, P., J. Clark, and J. Clark, Eds., *Wetland Functions and Values: The State of Our Understanding.* American Water Resources Association, Minneapolis, MN, pp.377-388.

Nyerges, T., 1993. Understanding the scope of GIS: Its relationship to environmental modeling. In Goodchild et al., 1993, pp.75-93.

Olson, M. 1992. The USEPA River Reach File 3: A National Hydrographic Database for GIS Analyses. Unpublished manuscript, USEPA, Las Vegas, NV.

Paul, J. and G. Morrison, 1995. Watershed stressors and Environmental Monitoring and Assessment Program estuarine indicators for south shore Rhode Island. In USEPA, 1995, pp.101-109.

Pearlstine, L., W. Kitchens, P. Latham, and R. Bartleson, 1993. Tide gate influences on a tidal marsh. *Water Resources Bulletin* 29:1009-1019.

Pickus, J. and M. Hewitt, 1992. Resource at risk: Analyzing sensitivity of groundwater to pesticides. *Geo Info Systems*, November-December, pp.50-56.

Pilon, P., P. Howarth, and R. Bullock, 1988. An enhanced classification approach to change detection in semi-arid environments. *Photogrammetric Engineering and Remote Sensing* 54:1709-1716.

Poiani, K. and B. Bedford, 1995. GIS-based nonpoint source pollution modeling: Considerations for wetlands. *Journal of Soil and Water Conservation* 50:613-619.

Pulich, W. and J. Hinson, 1995. Application of multitemporal Thematic Mapper data to change detection analysis of Texas coastal land cover. *Proceedings of the Third Thematic Conference on Remote Sensing of Marine and Coastal Environments*, ERIM, Ann Arbor, MI, Vol.2, pp.469-481.

Raabe, E. and R. Stumpf, 1995. Monitoring tidal marshes of Florida's Big Bend. *Proceedings of the Third Thematic Conference on Remote Sensing of Marine and Coastal Environments*, ERIM, Ann Arbor, MI, Vol.2, pp.483-494.

Ramsey, E., 1998. Radar remote sensing of wetlands. In Lunetta and Elvidge, 1998, pp. 211-243.

Ramsey, E., D. Chappell, D. Jacobs, S. Sapkota, and D. Baldwin, 1998. Resource management of forested wetlands: Hurricane impact and recovery mapped by combining Landsat TM and NOAA AVHRR data. *Photogrammetric Engineering and Remote Sensing* 64:733-738.

Ramsey, E. and J. Jensen, 1990. The derivation of water volume reflectances from airborne MSS data using in situ water volume reflectances, and a combined optimization technique and radiation transfer model. *International Journal of Remote Sensing* 11:979-998.

Ramsey, E. and J. Jensen, 1995. Modelling mangrove canopy reflectance using a light interaction model and an optimization technique. In Lyon and McCarthy, 1995, pp.61-75.

Ramsey, E. and J. Jensen, 1996. Remote sensing of mangrove wetlands: Relating canopy spectra to site-specific data. *Photogrammetric Engineering and Remote Sensing* 62:939-948.

Reed, P., 1988. National List of Plant Species that Occur in Wetlands: National Summary. U.S. Department of Interior, Washington, D.C.

Remillard, M. and R. Welch, 1992. GIS technologies for aquatic macrophyte studies: I. database development and changes in the aquatic environment. *Landscape Ecology* 7:151-162.

Remillard, M. and R. Welch, 1993. GIS technologies for aquatic macrophyte studies: Modeling applications. *Landscape Ecology* 8:163-175.

Rhodes, H. and D. Myers, 1993. GRASS used in the geostatistical analysis of lakewater data from the Eastern Lake Survey, phase I. In Goodchild et al., 1993, pp.438-446.

Richards, C. and G. Host, 1994. Using GIS to examine the influence of land use on stream habitats and biota. *Water Resources Bulletin* 30:729-738.

Richards, C., L. Johnson, and G. Host, 1995. Using GIS to examine linkages between landscapes and stream ecosystems. In USEPA, 1995, pp.131-141.

Richardson, A. and A. Milne, 1983. Mapping fire burns and vegetation regeneration using principal components analysis. *Proceedings of the IGARSS '83*, San Francisco, CA, pp.51-56.

# Bibliography

Ritchie, J. and C. Cooper, 1991. An algorithm for estimating surface suspended sediment concentration with Landsat MSS digital data. *Water Resources Bulletin* 27:373-379.

Robinson, G., R. Holt, M. Gaines, S. Hambur, M. Johnson, H. Fitch, and E. Martinko, 1992. Diverse and contrasting effects of habitat fragmentation. *Science* 257:524-526.

Roller, N., 1977. Remote Sensing of Wetlands. Technical Report No. 193400-14-T, Environmental Research Institute of Michigan, Ann Arbor, MI.

Salvesen, D., 1990. *Wetlands: Mitigating and Regulating Development Impacts.* Urban Land Institute, Washington, D.C.

Sample, V., 1994. *Remote Sensing and GIS in Ecosystem Management.* Island Press, Novato, CA.

Samuels, W., 1993. LAKEMAP: A 2-D and 3-D mapping system for visualizing water quality data in lakes. *Water Resources Bulletin* 29:917-922.

Schaal, G., 1995. Methods Used in the Ohio Wetland Inventory. Unpublished manuscript, Ohio Department of Natural Resources, Columbus, OH.

Scherz, J., 1977. Assessment of Aquatic Environment by Remote Sensing. Institute of Environmental Studies, University of Wisconsin, Madison, WI.

Schoolmaster, F. and P. Marr, 1992. Geographic Information Systems as a tool in water use data management. *Water Resources Bulletin* 28:331-336.

Scieszka, M., 1990. The Digital Wetlands Data Base for the U.S. Great Lakes Shoreline. Federal Coastal Wetland Mapping Programs, Biology Report 90 (18), U.S. Fish and Wildlife Service, Washington, D.C., pp.159-172.

Scott, J., F. Davis, B. Csulti, R. Noss, B. Butterfield, C. Groves, H. Anderson, S. Caicco, F. D'Erchia, T. Edwards, J. Ulliman, and R. Wright, 1993. Gap analysis: A geographic approach to protection of biological diversity. *Wildlife Monographs* 123:1-41.

Senay, G., A. Ward, J. Lyon, N. Fausey, S. Nokes, and L. Brown, 2000. The relations between spectral data and water in a crop production environment. *International Journal of Remote Sensing* 21:1897-1910.

Shaw, D., D. Field, T. Holm, M. Jennings, J. Sturdevant, G. Thelin, and L. Worthy, 1993. An innovative partnership for national environmental assessment. *Proceedings of the 12th Pecora Symposium*, Sioux Falls, SD, p.339.

Sherin, A. and K. Edwardson, 1995. A coastal information system for the Atlantic provinces of Canada. *Proceedings of the Third Thematic Conference on Remote Sensing of Marine and Coastal Environments*, ERIM, Ann Arbor, MI, Vol.2, pp.401-413.

Sinclair, R., M. Graves, and J. Stoll, 1990. Satellite Data and GIS Technology Applications to Wetlands Mapping. Federal Coastal Wetland Mapping Programs, Biology Report 90 (18), U.S. Fish and Wildlife Service, Washington, D.C., pp.151-157.

Singh, A., 1989. Digital change detection techniques using remotely-sensed data. *International Journal of Remote Sensing* 10:989-1003.

Slonecker, T., J. Owecke, L. Mata, and L. Fisher, 1992. GPS: Great gains in the great outdoors. *GPS World*, September, pp.25-34.

Soils Conservation Service, 1992. State Soils Geographic Database (STATSGO) User's Guide. Soil Conservation Service Publication, No. 1492.

Star, J. and J. Estes, 1990. *Geographic Information Systems: An Introduction*. Prentice Hall, Englewood Cliffs, NJ.

Stoms, D., 1992. Effects of habitat map generation in biodiversity assessment. *Photogrammetric Engineering and Remote Sensing* 58:1587-1591.

Story, M. and R. Congalton, 1986. Accuracy assessment: A user's perspective. *Photogrammetric Engineering and Remote Sensing* 52:397-399.

Sutter, L. and J. Wuenscher, 1995. Wetlands mapping and assessment in coastal North Carolina: A GIS-based approach. In USEPA, 1995, pp.199-212.

Tappan, G., D. Tyler, M. Wehde, and D. Moore, 1992. Monitoring rangeland dynamics in Senegal with Advanced Very High Resolution Radiometer data. *Geocarto International* 1:87-98.

Thapa, K. and J. Bossler, 1992. Accuracy of spatial data used in Geographic Information Systems. *Photogrammetric Engineering and Remote Sensing* 58:835-841.

Thenkabail, P. and C. Nolte, 1995. Mapping and Characterizing Inland Valley Agroccosystcms of West and Central Africa, A Methodology Integrating Remote Sensing, GPS, and Ground-Truth Data in a GIS Framework. Resource and Crop Management Research Monograph No.16, International Institute for Tropical Agriculture, Ibadan, Nigeria.

# Bibliography

Thenkabail, P., C. Nolte, and J. Lyon, 2000. Remote sensing and GIS modeling for selection of a benchmark research area in the inland valley agroecosystems of west and central Africa. *Photogrammetric Engineering and Remote Sensing* 66:755-768.

Thenkabail, P., A. Ward, J. Lyon, and C. Merry, 1994. Thematic Mapper vegetation indices for determining soybean and corn growth parameters. *Photogrammetric Engineering and Remote Sensing* 60:437-442.

Thenkabail, P., A. Ward, J. Lyon, and P. Van Deventer, 1992. Landsat Thematic Mapper data (TM) indices for evaluating management and growth characteristics of soybeans and corn. *Transaction of the American Society of Agricultural Engineers* 35:1441-1448.

Thompson, G. and R. Gauthier, 1990. Development of a GIS for the U.S. Great Lakes Shoreline. Unpublished manuscript, Detroit District, U.S. Army Corps of Engineers.

Toll, D., J. Royal, and J. Davis, 1980. Urban and regional land use change detected by using Landsat data. *Proceedings of the Fall Technical Meeting of the American Society of Photogrammetry, RS-E1-17.*

Townshend, J. and C. Justice, 1986. Analysis of the dynamics of African vegetation using the normalized difference vegetation index. *International Journal of Remote Sensing* 7:1435-1445.

Troge, M., 1995. A GIS strategy for lake management issues. In USEPA, 1995, pp.261-265.

U.S. Army Corps of Engineers, 1987. Corps of Engineers Wetlands Delineation Manual. Technical Report Y-87-1, Department of the Army, Washington, D.C.

U.S. Army Corps of Engineers, 1988. Draft Environmental Impact Statement, Supplement II to the Final Environmental Impact Statement, Operations, Maintenance, and Minor Improvements of the Federal Facilities at Sault Ste. Marie, Michigan (July, 1977). Detroit, MI.

U.S. Army Corps of Engineers, 1993. Photogrammetric Mapping. Engineering Manual, Washington, D.C.

U.S. Department of Agriculture, 1962. Soil Survey Manual. Soil Conservation Service, Washington, D.C.

U.S. Department of Agriculture, 1975. Soil Taxonomy, A Basic System of Soil Classification for Making and Interpreting Soil Surveys. U.S. Soil Conservation Service, Washington, D.C.

U.S. Department of Agriculture, 1991. Hydric Soils of the United States. Miscellaneous Publication 1491, Soil Conservation Service, Washington, D.C.

U.S. Environmental Protection Agency, 1991. Federal Manual for Identifying and Delineating Jurisdictional Wetlands. *Federal Register*, August 14.

U.S. Environmental Protection Agency, 1995. National Conference on Environmental Problem-Solving with Geographic Information Systems. Seminar Publication, EPA/625/R-95/004.

Van Deventer, P., A. Ward, and J. Lyon, 1997. Using Thematic Mapper data to identify contrasting soil plains and tillage practices. *Photogrammetric Engineering and Remote Sensing* 63:87-93.

Van Sickle, J., 1996. *GPS for Land Surveyors*. Ann Arbor Press, Chelsea, MI.

Vieux, B. and S. Needham, 1992. Nonpoint-pollution model sensitivity to grid-cell size. *Journal of Water Resources, Planning and Management* 119:141-157.

Walbridge, M., 1993. Functions and values of forested wetlands in the southern United States. *Journal of Forestry* 91:15-19.

Ward, A. and W. Elliot, 1995. *Environmental Hydrology*. Lewis Publishers, Boca Raton, FL.

Wayland, R., 1995. The Clinton administration's perspective on wetlands protection. *Journal of Soil and Water Conservation* 50:581-584.

Weismiller, R., S. Kristoof, D. Scholz, P. Anuta, and S. Momen, 1977. Change detection in coastal zone environments. *Photogrammetric Engineering and Remote Sensing* 43:1533-1539.

Welch, R., M. Remillard, and J. Alberts, 1992. Integration of GPS, remote sensing, and GIS techniques for coastal resource management. *Photogrammetric Engineering and Remote Sensing* 58:1571-1578.

Welch, R., M. Remillard, and R. Slack, 1988. Remote sensing and Geographic Information System techniques for aquatic resource evaluation. *Photogrammetric Engineering and Remote Sensing* 54:177-185.

White, D. and S. Fennessy, 1996. The Cuyahoga Watershed Demonstration Project for the Identification of Wetland Restoration Sites. Unpublished manuscript, Ohio Environmental Protection Agency.

# Bibliography

White, D., R. Smith, C. Richard, B. Alexander, and K. Robinson, 1992. A spatial model to aggregate point-source and nonpoint-source water quality data for large areas. *Computer and Geoscience* 18:1055-1073.

Wilen, B., 1990. The U.S. Fish and Wildlife Service's National Wetlands Inventory. Federal Coastal Wetland Mapping Programs, Biology Report 90 (18), U.S. Fish and Wildlife Service, Washington, D.C., pp.9-20.

Williams, D. and J. Lyon, 1991. Use of a Geographical Information System data base to measure and evaluate wetland changes in the St. Marys River, Michigan. *Hydrobiologia* 219:83-95.

Williams, D. and J. Lyon, 1997. Historical aerial photographs and a GIS to determine the effects of long-term water levels on wetlands along the St. Marys River, Michigan. *Aquatic Botany* 58:363-378.

Wood, E., M. Sivapalan, K. Beven, and L. Band, 1988. Effects of spatial variability and scale with implications to hydrologic modeling. *Journal of Hydrology* 102:29-47.

Wu, S., 1989. Multipolarization P-, L-, and C-band radar for coastal zone mapping: the Louisiana example. *Proceedings of the 1989 Fall Convention of ACSM/ASPRS*, Cleveland, OH, pp.178-183.

Yi, G., D. Risley, M. Koneff, and C. Davis, 1994. Development of Ohio's GIS-based wetlands inventory. *Journal of Soil and Water Conservation* 49:23-28.

Young, R. and T. Dahl, 1995. Locating wetland loss "hot spots" using GIS. *Professional Surveyor* 15:29-31.

Yuan, D., C. Elvidge, and R. Lunetta, 1998. Survey of multispectral methods for land cover change analysis. In Lunetta, R. and C. Elvidge, 1998, pp. 21-39.

# *Index*

**A**

accuracy assessments
    GIS data  105-107
    land cover products  22, 85-87
Advanced Very High Resolution Radiometer  102
aerial imagery  15i, 35i, 56i, 60i
    applications  20, 22,
    change detection  70-73
    elements of,  57-60
    film choices  22, 40
    historical  63-67, 96-97
    interpreting  60-62
    land cover products  70
    soil appearance  67-68
    vegetation appearance  67
    water resources  68-69
anaerobic soil  5i, 9-10, 24
Anderson system, USGS  20, 85

**B**

black & white film  22, 40
boundaries
    marking by GPS  38
    surveying  35-38
Brown system  85

**C**

change detection
    current approaches  74-78
    data processing  80-81
    identifying change  40-43, 70-73
    NALC project  79-80
    principal components analysis  82-83
    vegetative indices  81-82
classification systems (wetland)  20, 69, 85, 96
coastal wetlands  4i, 7i, 72
    trends  13-14
    emergent wetlands  14
    and GIS data  97-100
    influences upon  9, 41i
    lacustrine wetlands and GIS data  100-101
    plants and  16
Coastal Zone Management  99
color, analysis of  58
color infrared (CIR) film  22, 40
construction activity, remote detection of  40, 43-44
continental data sets  15
Cowardin system  69, 85

**D**

data processing  55, 80-81
data sources  92-95. *See also* map sources
data storage  11, 89-90
delineation boundaries
    marking methods  35-38
    *Practical Handbook for Wetland Identification and Delineation*  70
    *Wetlands Delineation Manual*  23
destriping  78
Differential GPS  38

Digital Elevation Models (DEM) 93-94
Digital Line Graphs (DLG) 17, 93
digital soils data sets 94
direct measurements 11
Ducks Unlimited 17

**E**

Electronic Distance Measurement (EDM) 38
emergent wetlands 14i, 101
erosion 50
estuary management 100

**F**

facultative upland plants (FACU) 25
facultative wetland plants (FACW) 25
facultative wetland species (FAC) 25
feature elements
    associations between 60
    descriptions of 57-60
field evaluation
    advanced analysis 31-34
    routine analysis 29-30
    surveying 35-38
film 22, 40, 58, 62
flooding 6i, 30i
    remote detection 48, 49
    surface indicators 27
foliage conditions
    appearance on film 40
    film choice for 22
    grey tone appearance 58

**G**

GAP program 51
general wetlands 9, 19
    distinct from jurisdictional 19-20
GIS (Geographic Information Systems). *See also* remote sensor data
    and aerial photo analysis 96-97
    applications 17, 92-95
    change detection procedures 42-43
    data storage 11, 89-90
    ground control points 37, 92-95
    lake related wetlands 100-101

    marine coastal wetland analysis 97-100
    models 90-91
    quality assessment of 91
    and remote sensor data 97
    river and lake management 103-104
    water quality evaluations 101-103
Global Land Information System (GLS) 80
Global Positioning Systems (GPS) 38
Great Lakes. *See* Laurentian Great Lakes
grey tones, analysis of 57-58
ground control points (GCP) 37
growing season
    definition 32
    and temporary wetlands 9

**H**

hazardous waste 53
hydric soils 24-25, 26
    field evaluation methods 31-32
    identifying characteristics 26-27
    USDA Hydric Soils List 27
hydrological analysis 9
    criterion satisfaction 31-32
    hydrological resource units (HRU) 104
    methods 27-29
    silt or clay deposits 24i, 50
    using large-scale maps 21

**I**

image elements 57-60
image enhancement
    change detection techniques 76-78
    misregistration errors 81
indirect measurements 11
interpretation 60-62
    definition 57-58

**J**

jurisdictional wetland 9
    determination criteria 23, 24
    distinct from general 19-20

**L**

lake storage 14, 96-97
land cover products 70, 83, 84-87

land development
  permits  35
  remote detection of  43-44
  site selection  52-53
  wetland risk  8
landscape characterization, definition  1
landscape ecology, definition  1
Landstat Multispectral Scanner (MSS) data  75, 78, 79
Landstat Thematic Mapper (TM) data  17, 67, 78
large area assessments  21-22
large-scale maps
  from aerial photos  86i
  common scale of  20-21
laser measurements  38
Laurentian Great Lakes  13i, 41i, 45, 96-97
local wetland inventories  22-23

## M

management  5, 10, 42-43
mangrove wetlands  16, 99
*1987 Manual,* USACE  29, 30, 31
mapping methods  19-38. *See also* wetland inventory
  land cover maps  70, 73, 85
mapping projects, goals  19
map sources  92-94
  human related activities  92
  large-scale maps  20-21
  navigable waterways  29, 92
  soil maps  26-27
  wetland maps  20-21, 92
methane gas  31, 47
models (GIS)  90-91
monitoring  43-46, 45-46
multiple date data  13-17
Munsell color charts  29, 31, 31i, 32

## N

NALC (North American Landscape Characterization) project  75, 79-80
National Archives and Records Service (NARS)  92

*National List of Plant Species that Occur in Wetlands: National Summary*  25, 32
National Oceanic and Atmospheric Administration (NOAA)  92, 102
National Wetland Inventory (NWI)  17, 21
Natural Resource Conservation Service (NRCS)  26
Nature Conservancy, The  17
navigable waterways  29, 92
NDVI (normalized vegetation index)  78-79
nitrogen  47
NOAA Coastwatch Program  102
nonpoint source pollution  100

## O

obligatory wetland plants (OBL)  4i, 25
Ohio Capability Analysis Program (OCAP)  17
Ohio Sea Grant  103
Ohio Wetlands Inventory  17

## P

patterns, analysis of  59
permits  35
photogrammetric calculations  37, 62
photointerpretation, definition  57
plants. *See* wetland plants
postcategorization methods  74, 83
  disadvantages of  77-78
potential jurisdictional wetlands  9, 19
*Practical Handbook for Wetland Identification and Delineation*  70
Preliminary Land Cover products  84-85
preprocessing of satellite data  78-79
principal components analysis  82-83
principal components data sets  82
public awareness
  common questions  70
  general perceptions  3, 5, 6i
  heightening  19, 47-48

## Q

quality assurance/quality control  105-107

## R

Reed, P. 25, 32
regional assessments 21-22
    advanced field methods 31-34
    DGPS measured boundaries 38
regional data sets 15
regulatory jurisdictional wetlands 23
remote sensor data. *See also* change detection
    accuracy assessments 85-87
    cloud effects 81, 82
    historical 63-67
    image enhancement 76-78, 81
    interpretation 16i, 60-62
    land cover products 70
    limitations 55
    photointerpretation 57-60
    postcategorization 74, 77-78, 83
    preprocessing of 78-79
    resolution and spectral bandwidth of 15, 80
    training set development 83-85
    vegetation and soil identification 67-68
    water resources identification 68-69
    wildlife habitat evaluations 87
remote sensor technology, range of choices 56
resolution, of sensor data 15
retention ponds 47i, 52i
risk assessment
    determining 2-3
    human related stressors 52-53, 92
    natural stressors 48-51, 99-100
riverine wetlands 9, 42i, 56i
    aerial view of 59i
    analysis with GIS data 103-104
    remote sensing of 49
runoff, seasonal 42i

## S

Salicornia wetlands 16
Sanborn Maps 92
sediments 62, 101-102
seiches 14, 41i
shapes, analysis of 58
sizes, analysis of 58
small area assessments 22-23
Soil Conservation Service (SCS) 26
Soil Survey documents (USDA) 20, 26
soil testing 31-32
soil types. *See also* hydric soils
    anaerobic 5i, 9-10
    digital soils data 94
    identifying 26-27, 27-29
    photointerpretation of 43-44, 58, 67-68
Spartina wetlands 16
spectral bandwidth 15, 80
spectral reflectance 60, 62
STATSGO 94
submergent wetlands 13i, 62, 98i, 101
subsurface hydrological test 27, 29
sulfur 31
surveying 35-38

## T

temporary wetlands 6i, 7i, 50i
    definition 9
    preservation of 53i
textures, analysis of 58
thematic mapping 69
thermal sensor data 102
tidal effects 2
    cold area wetlands 8i
    multiple date data 13-14
TIGER data 92
trace gasses 31, 47
training set
    categorization 84-85
    development 83-84
trends, identifying 13

## U

Universal Transverse Mercator (UTM) 38
unsupervised training sets 83-84
U.S. Army Corps of Engineers
    map source 29
    wetland definition 8-9

U.S. Geological Survey (USGS)  92
USACE HEP  87
USDA Hydric Soils List  27
USDA-NRCS  27
*USDA-Soil Taxonomy, 1975*  29
USGS EROS Data Center  79, 80

**V**

variables (in studies)
    control of  10
    GIS  95i, 96
vegetation. *See also* wetland plants
    image interpretation  58, 67
    leaf-on/leaf-off conditions  21, 22, 81, 87
vegetative indices  81-82

**W**

water conditions
    risk assessment  48, 49, 50, 56
    sediment concentrations  101-102
    surface indicators  27-28
    temperature data  102
water quality assessments  101-103
water resources, photointerpretation of  16i, 68-69
water saturation, soil response  24
water storage
    of the Great Lakes  96-97
    lake system wetlands  14, 101
wetland plants  4i, 6i
    adaptations  8-9, 24-25
    criterion satisfaction  25, 32
    field methods  32
    identification and frequency  25-26
wetlands
    definition  8-9
    federal criteria  9, 23
*Wetlands Delineation Manual*  23
wetlands inventory  17
    assessing criteria fulfillment  24-29
    local procedure  22-23
    purpose for  10
    regional procedure  21-22
wildlife habitat
    data sources  15, 51
    management  46i
    quality evaluations  87